J. J. Buckley

Fuzzy Probabilities and Fuzzy Sets for Web Planning

Springer

Berlin
Heidelberg
New York
Hong Kong
London
Milano
Paris
Tokyo

Studies in Fuzziness and Soft Computing, Volume 135

http://www.springer.de/cgi-bin/search_book.pl?series=2941

Editor-in-chief
Prof. Janusz Kacprzyk
Systems Research Institute
Polish Academy of Sciences
ul. Newelska 6
01-447 Warsaw
Poland
E-mail: kacprzyk@ibspan.waw.pl

James J. Buckley

Fuzzy Probabilities and
Fuzzy Sets for Web Planning

Springer

Prof. James J. Buckley
Mathematics Department
University of Alabama at Birmingham
Birmingham, AL 35294
USA
E-mail: buckley@math.uab.edu

ISSN 1434-9922
ISBN 3-540-00473-4 Springer-Verlag Berlin Heidelberg New York

Library of Congress Cataloging-in-Publication-Data applied for

A catalog record for this book is available from the Library of Congress.

Bibliographic information published by Die Deutsche Bibliothek
Die Deutsche Bibliothek lists this publication in the Deutsche Nationalbibliographie;
detailed bibliographic data is available in the internet at <http://dnb.ddb.de>.

Springer-Verlag Berlin Heidelberg New York
a member of BertelsmannSpringer Science+Business Media GmbH
http://www.springer.de

© Springer-Verlag Berlin Heidelberg 2004
Printed in Germany

Typesetting: Camera-ready by author
Cover design: E. Kirchner, Springer-Verlag, Heidelberg
Printed on acid free paper   62/3020/M - 5 4 3 2 1 0

To Julianne and Helen.

Contents

Chapter 1

Introduction

1.1 Introduction

This book is written in five major divisions. The first part is the introductory chapters consisting of Chapters 1-3. In part two, Chapters 4-10, we use fuzzy probabilities to model a fuzzy queuing system . We switch to employing fuzzy arrival rates and fuzzy service rates to model the fuzzy queuing system in part three in Chapters 11 and 12. Optimization models comprise part four in Chapters 13-17. The final part has a brief summary and suggestions for future research in Chapter 18, and a summary of our numerical methods for calculating fuzzy probabilities, values of objective functions in fuzzy optimization, etc., is in Chapter 19.

First we need to be familiar with fuzzy sets. All you need to know about fuzzy sets for this book comprises Chapter 2. Two other items relating to fuzzy sets, needed in Chapters 13-17, are also in Chapter 2: (1) how we plan to handle the maximum/minimum of a fuzzy set; and (2) how we will rank a finite collection of fuzzy numbers from smallest to largest.

1.2 Fuzzy Probabilities

In this part of the book we will apply our new method of using fuzzy numbers to model uncertain probabilities. Details of our new approach to handing uncertain probabilities is in [1] and the book [2]. Two parts of our new method of modeling uncertain probabilities used in this book are: (1) using a set of confidence intervals to construct a fuzzy number for an uncertain probability; and (2) restricted fuzzy arithmetic. We show how fuzzy probabilities can be constructed from a set of confidence intervals in Chapter 3. Restricted fuzzy arithmetic, first proposed in the papers ([3]-[7]), is also discussed in Chapter 3. The method of finding fuzzy probabilities, using restricted fuzzy arithmetic, usually involves finding the maximum, and minimum, of a linear or non-linear function, subject to linear constraints. These types of computations are used throughout the book and are first presented in Chapter 3. Further discussion about these calculations are in Chapter 19.

Restricted fuzzy arithmetic, extended to restricted fuzzy matrix multiplication, is used in fuzzy Markov chains in Chapter 4. We model a fuzzy queuing system as a fuzzy finite, (usually) regular, Markov chain in Chapter 5. Using the results from Chapter 4 we obtain the fuzzy steady state probabilities in the fuzzy queuing system.

All the computations needed to get to the fuzzy numbers for system performance are described in the Chapters 6-10. There are three basic steps in the computations, discussed in Chapters 6-10, starting with only one server in Chapter 6. All our fuzzy queuing systems will have finite capacity denoted by M. Chapter 7 is a numerical example for Chapter 6. Chapter 8 considers two servers followed by a numerical example in Chapter 9. We end with more than three servers in Chapter 10. The spread on the fuzzy numbers for system performance shows the uncertainties in their values. The base of these fuzzy numbers is like a 99% confidence interval.

1.3   Fuzzy Arrival/Service Rates

The other method of modeling a fuzzy queuing system is through the arrival rate and the service rate. We start with gathering some data to estimate these rates and show, in Chapter 3, how we can use confidence intervals to construct fuzzy numbers for these rates. Starting with fuzzy numbers for the arrival rate and service rate we find the fuzzy steady state probabilities and then the fuzzy numbers for system performance all in Chapter 11. A numerical example is presented in Chapter 12. Also in this modeling method, the spread of the fuzzy numbers for system performance gives the uncertainty in their value.

1.4   Optimization Models

Optimization models are discussed in Chapters 13-17. We start with ignoring revenues/costs in Chapter 13. In Chapter 13 we are mainly concerned with minimizing $\overline{R}$, the fuzzy number representing average response time for the system, and in maximizing $\overline{U}$, the fuzzy number for server utilization. Two solution methods are given. Costs/revenues are added in Chapter 14 where we consider server costs, queue costs,..., and revenues to maximize the fuzzy number for profit.

Two phenomenon identified with web sites, "burstiness" and " long tailed distributions" are in Chapters 15 and 16, respectively. We model burstiness with two types of arrivals and we present a different optimization model involving "ideal" points, being fuzzy numbers for $\overline{R}$, to find an optimal mix of the variables. The other phenomena, long tailed distributions, is modeled as two classes of customers producing two arrival patterns and two service times.

We have looked at different types of sub-optimization models in Chapters 13-16 and in Chapter 17 we put it all together into a global optimization plan. In each case, except in Chapter 17, we first present an analytical model and then a graphical model based on the ranking of the fuzzy sets from smallest to largest.

Chapters 13-17 have been written so that the optimization models based on fuzzy probabilities (always presented first) and the optimization models using fuzzy arrival/sevice rates (discussed second) , can be studied independently. This means that the development of the models is presented twice, once for fuzzy probabilities and again for fuzzy arrival/service rates. So, if you go through all of one of these chapters some material will be repeated when you get to fuzzy arrival/service rates.

1.5 Notation

It is difficult, in a book with a lot of mathematics, to achieve a uniform notation without having to introduce many new specialized symbols. Our basic notation is presented in Chapter 2. What we have done is to have a uniform notation within each section. What this means is that we may use the letters "a" and "b" to represent a closed interval $[a, b]$ in one section but they could stand for parameters in an optimization model in another section.

We will have the following uniform notation throughout the book:

(1) we place a "bar" over a letter to denote a fuzzy set ($\overline{A}$, $\overline{B}$, etc.), and all our fuzzy sets will be fuzzy subsets of the real numbers;

(2) an alpha-cut is always denoted by "α", see Chapter 2;

(3) fuzzy functions are denoted as $\overline{F}$, $\overline{G}$, etc., in Chapter 2;

All fuzzy arithmetic is performed using α-cuts and interval arithmetic and not by using the extension principle (Chapter 2).

The term "crisp" means not fuzzy. A crisp set is a regular set and a crisp number is a real number. There is a potential problem with the symbol "$\leq$". It usually means "fuzzy subset" as $\overline{A} \leq \overline{B}$ stands for $\overline{A}$ is a fuzzy subset of $\overline{B}$ (defined in Chapter 2). However, also in Chapter 2, $\overline{A} \leq \overline{B}$ means that fuzzy set $\overline{A}$ is less than or equal to fuzzy set $\overline{B}$. The meaning of the symbol "$\leq$" should be clear from its use, but we shall point out when it will mean $\overline{A}$ is less that or equal to $\overline{B}$. Also, in Chapter 3 $\overline{X}$ will be the mean of a random sample, not a fuzzy set, and we explicitly point this out when it arises in that chapter.

There is another possible notational problem when we use the same symbol with, and without, a "bar" on top in the same section. In Chapters 13-17 we will have λ (and μ) and $\overline{\lambda}$ (and $\overline{\mu}$) but they represent different objects: (1) λ (μ) is a real number; and (2) $\overline{\lambda}$ ($\overline{\mu}$) stands for a fuzzy number (defined in Chapter 2).

Prerequisites are a basic knowledge of crisp probability theory and elementary calculus. We do take a few derivatives. There are numerous text books on probability theory, so there no need to give references for probability theory.

The recommended chapter flow is as follows: (1) using fuzzy probabilities

$$\{1,2,3\} \rightarrow \{4-10\} \rightarrow first\,part\,\{13-17\} \rightarrow \{18,19\}; \qquad (1.1)$$

and (2) using fuzzy arrival/service rates

$$\{1,2,3\} \rightarrow \{11,12\} \rightarrow second\,part\,\{13-17\} \rightarrow \{18,19\}. \qquad (1.2)$$

1.6   References

1. J.J.Buckley and E.Eslami: Uncertain Probabilities I: The Discrete Case, Soft Computing. To appear.

2. J.J.Buckley: Fuzzy Probabilities: New Approach and Applications, Physics-Verlag, Heidelberg, Germany, 2003.

3. G.J.Klir: Fuzzy Arithmetic with Requisite Constraints, Fuzzy Sets and Systems, 91(1997), pp. 147-161.

4. G.J.Klir and J.A.Cooper: On Constrainted Fuzzy Arithmetic, Proc. 5th Int. IEEE Conf. on Fuzzy Systems, New Orleans, 1996, pp. 1285-1290.

5. G.J.Klir and Y.Pan: Constrained Fuzzy Arithmetic: Basic Questions and Some Answers, Soft Computing, 2(1998), pp. 100-108.

6. Y.Pan and B.Yuan: Baysian Inference of Fuzzy Probabilities, Int. J. General Systems, 26(1997), pp. 73-90.

7. Y.Pan and G.J.Klir: Bayesian Inference Based on Interval-Valued Prior Distributions and Likelihoods, J. of Intelligent and Fuzzy Systems, 5(1997), pp. 193-203.

Chapter 2

Fuzzy Sets

2.1 Introduction

In this chapter we have collected together the basic ideas from fuzzy sets and fuzzy functions needed for the book. Any reader familiar with fuzzy sets, fuzzy numbers, the extension principle, α-cuts, interval arithmetic, and fuzzy functions may go on and have a look at Sections 2.5 and 2.6. In Section 2.5 we discuss our method of handling the maximun/minimum of a fuzzy set and in Section 2.6 we present a method of ordering a finite set of fuzzy numbers from smallest to largest. Both Section 2.5 and 2.6 will be employed in Chapters 13-17. A good general reference for fuzzy sets and fuzzy logic is [4] and [17].

Our notation specifying a fuzzy set is to place a "bar" over a letter. So $\overline{A}$, $\overline{B}$,..., $\overline{X}$, $\overline{Y}$,..., $\overline{\alpha}$, $\overline{\beta}$,..., will all denote fuzzy sets.

2.2 Fuzzy Sets

If Ω is some set, then a fuzzy subset $\overline{A}$ of Ω is defined by its membership function, written $\overline{A}(x)$, which produces values in $[0, 1]$ for all x in Ω. So, $\overline{A}(x)$ is a function mapping Ω into $[0, 1]$. If $\overline{A}(x_0) = 1$, then we say x_0 belongs to $\overline{A}$, if $\overline{A}(x_1) = 0$ we say x_1 does not belong to $\overline{A}$, and if $\overline{A}(x_2) = 0.6$ we say the membership value of x_2 in $\overline{A}$ is 0.6. When $\overline{A}(x)$ is always equal to one or zero we obtain a crisp (non-fuzzy) subset of Ω. For all fuzzy sets $\overline{B}$, $\overline{C}$,... we use $\overline{B}(x)$, $\overline{C}(x)$,... for the value of their membership function at x. Most of the fuzzy sets we will be using will be fuzzy numbers .

The term "crisp" will mean not fuzzy. A crisp set is a regular set. A crisp number is just a real number. A crisp matrix (vector) has real numbers as its elements. A crisp function maps real numbers (or real vectors) into real numbers. A crisp solution to a problem is a solution involving crisp sets, crisp numbers, crisp functions, etc.

2.2.1   Fuzzy Numbers

A general definition of fuzzy number may be found in ([4],[17]), however our fuzzy numbers will be almost always triangular (shaped), or trapezoidal (shaped), fuzzy numbers. A triangular fuzzy number $\overline{N}$ is defined by three numbers $a < b < c$ where the base of the triangle is the interval $[a, c]$ and its vertex is at $x = b$. Triangular fuzzy numbers will be written as $\overline{N} = (a/b/c)$. A triangular fuzzy number $\overline{N} = (1.2/2/2.4)$ is shown in Figure 2.1. We see that $\overline{N}(2) = 1$, $\overline{N}(1.6) = 0.5$, etc.

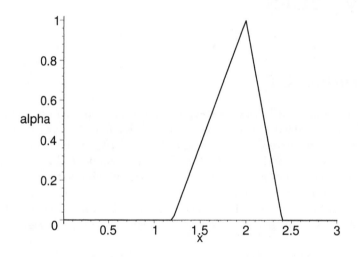

Figure 2.1: Triangular Fuzzy Number $\overline{N}$

A trapezoidal fuzzy number $\overline{M}$ is defined by four numbers $a < b < c < d$ where the base of the trapezoid is the interval $[a, d]$ and its top (where the membership equals one) is over $[b, c]$. We write $\overline{M} = (a/b, c/d)$ for trapezoidal fuzzy numbers. Figure 2.2 shows $\overline{M} = (1.2/2, 2.4/2.7)$.

A triangular shaped fuzzy number $\overline{P}$ is given in Figure 2.3. $\overline{P}$ is only partially specified by the three numbers 1.2, 2, 2.4 since the graph on $[1.2, 2]$, and $[2, 2.4]$, is not a straight line segment. To be a triangular shaped fuzzy number we require the graph to be continuous and: (1) monotonically increasing on $[1.2, 2]$; and (2) monotonically decreasing on $[2, 2.4]$. For triangular shaped fuzzy number $\overline{P}$ we use the notation $\overline{P} \approx (1.2/2/2.4)$ to show that it is partially defined by the three numbers 1.2, 2, and 2.4. If $\overline{P} \approx (1.2/2/2.4)$ we know its base is on the interval $[1.2, 2.4]$ with vertex (membership value one) at $x = 2$. Similarly we define trapezoidal shaped fuzzy number $\overline{Q} \approx (1.2/2, 2.4/2.7)$ whose base is $[1.2, 2.7]$ and top is over the interval $[2, 2.4]$. The graph of $\overline{Q}$ is similar to $\overline{M}$ in Figure 2.2 but it has

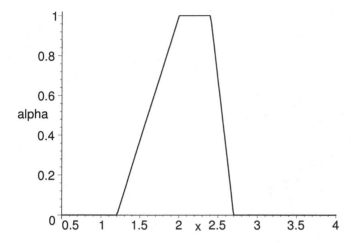

Figure 2.2: Trapezoidal Fuzzy Number $\overline{M}$

continuous curves for its sides.

Although we will be using triangular (shaped) and trapezoidal (shaped) fuzzy numbers throughout the book, many results can be extended to more general fuzzy numbers, but we shall be content to work with only these special fuzzy numbers.

We will be using fuzzy numbers in this book to describe uncertainty. For example, in Chapter 3 a fuzzy probability can be a triangular shaped fuzzy number, it could also be a trapezoidal shaped fuzzy number. Also, in Chapter 3 a fuzzy arrival (service) rate will be a triangular shaped fuzzy number. In Chapters 13-17 certain parameters in the optimization models are fuzzy numbers. In all these cases the uncertainty comes from estimating a value through a random sample, or from expert opinion.

2.2.2 Alpha-Cuts

Alpha-cuts are slices through a fuzzy set producing regular (non-fuzzy) sets. If $\overline{A}$ is a fuzzy subset of some set Ω, then an α-cut of $\overline{A}$, written $\overline{A}[\alpha]$ is defined as

$$\overline{A}[\alpha] = \{x \in \Omega | \overline{A}(x) \geq \alpha\}, \tag{2.1}$$

for all α, $0 < \alpha \leq 1$. The $\alpha = 0$ cut, or $\overline{A}[0]$, must be defined separately.

Let $\overline{N}$ be the fuzzy number in Figure 2.1. Then $\overline{N}[0] = [1.2, 2.4]$. Notice that using equation (2.1) to define $\overline{N}[0]$ would give $\overline{N}[0] =$ all the real numbers. Similarly, $\overline{M}[0] = [1.2, 2.7]$ from Figure 2.2 and in Figure 2.3 $\overline{P}[0] = [1.2, 2.4]$. For any fuzzy set $\overline{A}$, $\overline{A}[0]$ is called the support, or base, of $\overline{A}$. Many authors call the support of a fuzzy number the open interval

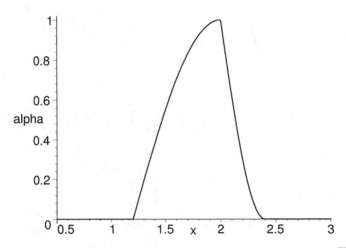

Figure 2.3: Triangular Shaped Fuzzy Number $\overline{P}$

(a, b) like the support of $\overline{N}$ in Figure 2.1 would then be $(1.2, 2.4)$. However in this book we use the closed interval $[a, b]$ for the support (base) of the fuzzy number.

The core of a fuzzy number is the set of values where the membership value equals one. If $\overline{N} = (a/b/c)$, or $\overline{N} \approx (a/b/c)$, then the core of $\overline{N}$ is the single point b. However, if $\overline{M} = (a/b, c/d)$, or $\overline{M} \approx (a/b, c/d)$, then the core of $\overline{M} = [b, c]$.

For any fuzzy number $\overline{Q}$ we know that $\overline{Q}[\alpha]$ is a closed, bounded, interval for $0 \leq \alpha \leq 1$. We will write this as

$$\overline{Q}[\alpha] = [q_1(\alpha), q_2(\alpha)], \tag{2.2}$$

where $q_1(\alpha)$ $(q_2(\alpha))$ will be an increasing (decreasing) function of α with $q_1(1) \leq q_2(1)$. If $\overline{Q}$ is a triangular shaped or a trapezoidal shaped fuzzy number then: (1) $q_1(\alpha)$ will be a continuous, monotonically increasing function of α in $[0, 1]$; (2) $q_2(\alpha)$ will be a continuous, monotonically decreasing function of α, $0 \leq \alpha \leq 1$; and (3) $q_1(1) = q_2(1)$ $(q_1(1) < q_2(1)$ for trapezoids). We sometimes check monotone increasing (decreasing) by showing that $dq_1(\alpha)/d\alpha > 0$ $(dq_2(\alpha)/d\alpha < 0)$ holds.

For the $\overline{N}$ in Figure 2.1 we obtain $\overline{N}[\alpha] = [n_1(\alpha), n_2(\alpha)]$, $n_1(\alpha) = 1.2 + 0.8\alpha$ and $n_2(\alpha) = 2.4 - 0.4\alpha$, $0 \leq \alpha \leq 1$. Similarly, $\overline{M}$ in Figure 2.2 has $\overline{M}[\alpha] = [m_1(\alpha), m_2(\alpha)]$, $m_1(\alpha) = 1.2 + 0.8\alpha$ and $m_2(\alpha) = 2.7 - 0.3\alpha$, $0 \leq \alpha \leq 1$. The equations for $n_i(\alpha)$ and $m_i(\alpha)$ are backwards. With the y-axis vertical and the x-axis horizontal the equation $n_1(\alpha) = 1.2 + 0.8\alpha$ means $x = 1.2 + 0.8y$, $0 \leq y \leq 1$. That is, the straight line segment from $(1.2, 0)$ to $(2, 1)$ in Figure 2.1 is given as x a function of y whereas it is usually stated as

y a function of x. This is how it will be done for all α-cuts of fuzzy numbers.

2.2.3 Inequalities

Let $\overline{N} = (a/b/c)$. We write $\overline{N} \geq \delta$, δ some real number, if $a \geq \delta$, $\overline{N} > \delta$ when $a > \delta$, $\overline{N} \leq \delta$ for $c \leq \delta$ and $\overline{N} < \delta$ if $c < \delta$. We use the same notation for triangular shaped and trapezoidal (shaped) fuzzy numbers whose support is the interval $[a, c]$.

If $\overline{A}$ and $\overline{B}$ are two fuzzy subsets of a set Ω, then $\overline{A} \leq \overline{B}$ means $\overline{A}(x) \leq \overline{B}(x)$ for all x in Ω, or $\overline{A}$ is a fuzzy subset of $\overline{B}$. $\overline{A} < \overline{B}$ holds when $\overline{A}(x) < \overline{B}(x)$, for all x. There is a potential problem with the symbol $\leq$. In some places in the book, for example see Section 2.6 and in Chapters 13-17, $\overline{M} \leq \overline{N}$, for fuzzy numbers $\overline{M}$ and $\overline{N}$, means that $\overline{M}$ is less than or equal to $\overline{N}$. It should be clear on how we use "$\leq$" as to which meaning is correct.

2.2.4 Discrete Fuzzy Sets

Let $\overline{A}$ be a fuzzy subset of Ω. If $\overline{A}(x)$ is not zero only at a finite number of x values in Ω, then $\overline{A}$ is called a discrete fuzzy set. Suppose $\overline{A}(x)$ is not zero only at x_1, x_2, x_3 and x_4 in Ω. Then we write the fuzzy set as

$$\overline{A} = \{\frac{\mu_1}{x_1}, \cdots, \frac{\mu_4}{x_4}\}, \tag{2.3}$$

where the μ_i are the membership values. That is, $\overline{A}(x_i) = \mu_i$, $1 \leq i \leq 4$, and $\overline{A}(x) = 0$ otherwise. We can have discrete fuzzy subsets of any space Ω. Notice that α-cuts of discrete fuzzy sets of $\mathbb{R}$, the set of real numbers, do not produce closed, bounded, intervals.

2.3 Fuzzy Arithmetic

If $\overline{A}$ and $\overline{B}$ are two fuzzy numbers we will need to add, subtract, multiply and divide them. There are two basic methods of computing $\overline{A} + \overline{B}$, $\overline{A} - \overline{B}$, etc. which are: (1) extension principle; and (2) α-cuts and interval arithmetic.

2.3.1 Extension Principle

Let $\overline{A}$ and $\overline{B}$ be two fuzzy numbers. If $\overline{A} + \overline{B} = \overline{C}$, then the membership function for $\overline{C}$ is defined as

$$\overline{C}(z) = \sup_{x,y}\{\min(\overline{A}(x), \overline{B}(y))|x + y = z\} . \tag{2.4}$$

If we set $\overline{C} = \overline{A} - \overline{B}$, then

$$\overline{C}(z) = \sup_{x,y}\{\min(\overline{A}(x), \overline{B}(y))|x - y = z\} . \tag{2.5}$$

Similarly, $\overline{C} = \overline{A} \cdot \overline{B}$, then

$$\overline{C}(z) = \sup_{x,y}\{\min(\overline{A}(x), \overline{B}(y))|x \cdot y = z\}, \tag{2.6}$$

and if $\overline{C} = \overline{A}/\overline{B}$,

$$\overline{C}(z) = \sup_{x,y}\{\min(\overline{A}(x), \overline{B}(y))|x/y = z\} . \tag{2.7}$$

In all cases $\overline{C}$ is also a fuzzy number [17]. We assume that zero does not belong to the support of $\overline{B}$ in $\overline{C} = \overline{A}/\overline{B}$. If $\overline{A}$ and $\overline{B}$ are triangular (trapezoidal) fuzzy numbers then so are $\overline{A} + \overline{B}$ and $\overline{A} - \overline{B}$, but $\overline{A} \cdot \overline{B}$ and $\overline{A}/\overline{B}$ will be triangular (trapezoidal) shaped fuzzy numbers.

We should mention something about the operator "sup" in equations (2.4)-(2.7). If Ω is a set of real numbers bounded above (there is a M so that $x \leq M$, for all x in Ω), then $\sup(\Omega)$ = the least upper bound for Ω. If Ω has a maximum member, then $\sup(\Omega) = \max(\Omega)$. For example, if $\Omega = [0,1)$, $\sup(\Omega) = 1$ but if $\Omega = [0,1]$, then $\sup(\Omega) = \max(\Omega) = 1$. The dual operator to "sup" is "inf". If Ω is bounded below (there is a M so that $M \leq x$ for all $x \in \Omega$), then $\inf(\Omega)$ = the greatest lower bound. For example, for $\Omega = (0,1]$ $\inf(\Omega) = 0$ but if $\Omega = [0,1]$, then $\inf(\Omega) = \min(\Omega) = 0$.

Obviously, given $\overline{A}$ and $\overline{B}$, equations (2.4)- (2.7) appear quite complicated to compute $\overline{A} + \overline{B}$, $\overline{A} - \overline{B}$, etc. So, we now present an equivalent procedure based on α-cuts and interval arithmetic. First, we present the basics of interval arithmetic.

2.3.2   Interval Arithmetic

We only give a brief introduction to interval arithmetic. For more information the reader is referred to ([18],[19]). Let $[a_1, b_1]$ and $[a_2, b_2]$ be two closed, bounded, intervals of real numbers. If $*$ denotes addition, subtraction, multiplication, or division, then $[a_1, b_1] * [a_2, b_2] = [\alpha, \beta]$ where

$$[\alpha, \beta] = \{a * b | a_1 \leq a \leq b_1, a_2 \leq b \leq b_2\} . \tag{2.8}$$

If $*$ is division, we must assume that zero does not belong to $[a_2, b_2]$. We may simplify equation (2.8) as follows:

$$[a_1, b_1] + [a_2, b_2] = [a_1 + a_2, b_1 + b_2], \tag{2.9}$$

$$[a_1, b_1] - [a_2, b_2] = [a_1 - b_2, b_1 - a_2], \tag{2.10}$$

$$[a_1, b_1] / [a_2, b_2] = [a_1, b_1] \cdot \left[\frac{1}{b_2}, \frac{1}{a_2}\right], \tag{2.11}$$

and

$$[a_1, b_1] \cdot [a_2, b_2] = [\alpha, \beta], \tag{2.12}$$

where

$$\alpha = \min\{a_1 a_2, a_1 b_2, b_1 a_2, b_1 b_2\}, \qquad (2.13)$$
$$\beta = \max\{a_1 a_2, a_1 b_2, b_1 a_2, b_1 b_2\} . \qquad (2.14)$$

Multiplication and division may be further simplified if we know that $a_1 > 0$ and $b_2 < 0$, or $b_1 > 0$ and $b_2 < 0$, etc. For example, if $a_1 \geq 0$ and $a_2 \geq 0$, then

$$[a_1, b_1] \cdot [a_2, b_2] = [a_1 a_2, b_1 b_2], \qquad (2.15)$$

and if $b_1 < 0$ but $a_2 \geq 0$, we see that

$$[a_1, b_1] \cdot [a_2, b_2] = [a_1 b_2, a_2 b_1] . \qquad (2.16)$$

Also, assuming $b_1 < 0$ and $b_2 < 0$ we get

$$[a_1, b_1] \cdot [a_2, b_2] = [b_1 b_2, a_1 a_2], \qquad (2.17)$$

but $a_1 \geq 0$, $b_2 < 0$ produces

$$[a_1, b_1] \cdot [a_2, b_2] = [a_2 b_1, b_2 a_1] . \qquad (2.18)$$

2.3.3 Fuzzy Arithmetic

Again we have two fuzzy numbers $\overline{A}$ and $\overline{B}$. We know α-cuts are closed, bounded, intervals so let $\overline{A}[\alpha] = [a_1(\alpha), a_2(\alpha)]$, $\overline{B}[\alpha] = [b_1(\alpha), b_2(\alpha)]$. Then if $\overline{C} = \overline{A} + \overline{B}$ we have

$$\overline{C}[\alpha] = \overline{A}[\alpha] + \overline{B}[\alpha] . \qquad (2.19)$$

We add the intervals using equation (2.9). Setting $\overline{C} = \overline{A} - \overline{B}$ we get

$$\overline{C}[\alpha] = \overline{A}[\alpha] - \overline{B}[\alpha], \qquad (2.20)$$

for all α in $[0, 1]$. Also

$$\overline{C}[\alpha] = \overline{A}[\alpha] \cdot \overline{B}[\alpha], \qquad (2.21)$$

for $\overline{C} = \overline{A} \cdot \overline{B}$ and

$$\overline{C}[\alpha] = \overline{A}[\alpha]/\overline{B}[\alpha], \qquad (2.22)$$

when $\overline{C} = \overline{A}/\overline{B}$, provided that zero does not belong to $\overline{B}[\alpha]$ for all α. This method is equivalent to the extension principle method of fuzzy arithmetic [17]. Obviously, this procedure, of α-cuts plus interval arithmetic, is more user (and computer) friendly.

Example 2.3.3.1

Let $\overline{A} = (-3/-2/-1)$ and $\overline{B} = (4/5/6)$. We determine $\overline{A} \cdot \overline{B}$ using α-cuts and interval arithmetic. We compute $\overline{A}[\alpha] = [-3 + \alpha, -1 - \alpha]$ and $\overline{B}[\alpha] = [4 + \alpha, 6 - \alpha]$. So, if $\overline{C} = \overline{A} \cdot \overline{B}$ we obtain $\overline{C}[\alpha] = [(\alpha - 3)(6 - \alpha), (-1 - \alpha)(4 + \alpha)]$, $0 \leq \alpha \leq 1$. The graph of $\overline{C}$ is shown in Figure 2.4.

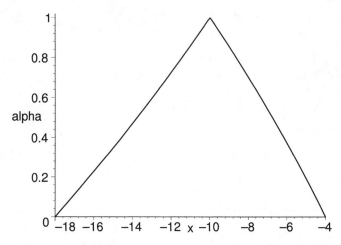

Figure 2.4: The Fuzzy Number $\overline{C} = \overline{A} \cdot \overline{B}$

2.4 Fuzzy Functions

In this book a fuzzy function is a mapping from fuzzy numbers into fuzzy numbers. We write $H(\overline{X}) = \overline{Z}$ for a fuzzy function with one independent variable $\overline{X}$. Usually $\overline{X}$ will be a triangular (trapezoidal) fuzzy number and then we usually obtain $\overline{Z}$ as a triangular (trapezoidal) shaped fuzzy number. For two independent variables we have $H(\overline{X}, \overline{Y}) = \overline{Z}$.

Where do these fuzzy functions come from? They are usually extensions of real-valued functions. Let $h : [a, b] \to \mathbb{R}$. This notation means $z = h(x)$ for x in $[a, b]$ and z a real number. One extends $h : [a, b] \to \mathbb{R}$ to $H(\overline{X}) = \overline{Z}$ in two ways: (1) the extension principle; or (2) using α-cuts and interval arithmetic.

2.4.1 Extension Principle

Any $h : [a, b] \to \mathbb{R}$ may be extended to $H(\overline{X}) = \overline{Z}$ as follows

$$\overline{Z}(z) = \sup_x \left\{ \, \overline{X}(x) \mid h(x) = z, \ a \le x \le b \, \right\} . \qquad (2.23)$$

Equation (2.23) defines the membership function of $\overline{Z}$ for any triangular (trapezoidal) fuzzy number $\overline{X}$ in $[a, b]$.

If h is continuous, then we have a way to find α-cuts of $\overline{Z}$. Let $\overline{Z}[\alpha] = [z_1(\alpha), z_2(\alpha)]$. Then [8]

$$z_1(\alpha) \ = \ \min\{ \, h(x) \mid x \in \overline{X}[\alpha] \, \}, \qquad (2.24)$$

$$z_2(\alpha) \ = \ \max\{ \, h(x) \mid x \in \overline{X}[\alpha] \, \}, \qquad (2.25)$$

for $0 \le \alpha \le 1$.

If we have two independent variables, then let $z = h(x, y)$ for x in $[a_1, b_1]$, y in $[a_2, b_2]$. We extend h to $H(\overline{X}, \overline{Y}) = \overline{Z}$ as

$$\overline{Z}(z) = \sup_{x,y} \left\{ \min \left(\overline{X}(x), \overline{Y}(y) \right) \mid h(x, y) = z \right\}, \qquad (2.26)$$

for $\overline{X}$ ($\overline{Y}$) a triangular or trapezoidal fuzzy number in $[a_1, b_1]$ ($[a_2, b_2]$). For α-cuts of $\overline{Z}$, assuming h is continuous, we have

$$z_1(\alpha) = \min\{ h(x, y) \mid x \in \overline{X}[\alpha], \ y \in \overline{Y}[\alpha] \}, \qquad (2.27)$$
$$z_2(\alpha) = \max\{ h(x, y) \mid x \in \overline{X}[\alpha], \ y \in \overline{Y}[\alpha] \}, \qquad (2.28)$$

$0 \le \alpha \le 1$. We use equations (2.24)- (2.25) and (2.27)-(2.28) throughout this book.

2.4.2 Alpha-Cuts and Interval Arithmetic

All the functions we usually use in engineering and science have a computer algorithm which, using a finite number of additions, subtractions, multiplications and divisions, can evaluate the function to required accuracy [7]. Such functions can be extended, using α-cuts and interval arithmetic, to fuzzy functions. Let $h : [a, b] \to \mathbb{R}$ be such a function. Then its extension $H(\overline{X}) = \overline{Z}$, $\overline{X}$ in $[a, b]$ is done, via interval arithmetic, in computing $h(\overline{X}[\alpha]) = \overline{Z}[\alpha]$, α in $[0, 1]$. We input the interval $\overline{X}[\alpha]$, perform the arithmetic operations needed to evaluate h on this interval, and obtain the interval $\overline{Z}[\alpha]$. Then put these α-cuts together to obtain the value $\overline{Z}$. The extension to more independent variables is straightforward.

For example, consider the fuzzy function

$$\overline{Z} = H(\overline{X}) = \frac{\overline{A}\, \overline{X} + \overline{B}}{\overline{C}\, \overline{X} + \overline{D}}, \qquad (2.29)$$

for triangular fuzzy numbers $\overline{A}, \overline{B}, \overline{C}, \overline{D}$ and triangular fuzzy number $\overline{X}$ in $[0, 10]$. We assume that $\overline{C} \ge 0$, $\overline{D} > 0$ so that $\overline{C}\, \overline{X} + \overline{D} > 0$. This would be the extension of

$$h(x_1, x_2, x_3, x_4, x) = \frac{x_1 x + x_2}{x_3 x + x_4}. \qquad (2.30)$$

We would substitute the intervals $\overline{A}[\alpha]$ for x_1, $\overline{B}[\alpha]$ for x_2, $\overline{C}[\alpha]$ for x_3, $\overline{D}[\alpha]$ for x_4 and $\overline{X}[\alpha]$ for x, do interval arithmetic, to obtain interval $\overline{Z}[\alpha]$ for $\overline{Z}$. Alternatively, the fuzzy function

$$\overline{Z} = H(\overline{X}) = \frac{2\overline{X} + 10}{3\overline{X} + 4}, \qquad (2.31)$$

would be the extension of

$$h(x) = \frac{2x + 10}{3x + 4}. \qquad (2.32)$$

2.4.3 Differences

Let $h : [a, b] \to \mathbb{R}$. Just for this subsection let us write $\overline{Z}^* = H(\overline{X})$ for the extension principle method of extending h to H for $\overline{X}$ in $[a, b]$. We denote $\overline{Z} = H(\overline{X})$ for the α-cut and interval arithmetic extension of h .

We know that $\overline{Z}$ can be different from $\overline{Z}^*$. But for basic fuzzy arithmetic in Section 2.2 the two methods give the same results. In the example below we show that for $h(x) = x(1 - x)$, x in $[0, 1]$, we can get $\overline{Z}^* \neq \overline{Z}$ for some $\overline{X}$ in $[0, 1]$. What is known ([8],[18]) is that for usual functions in science and engineering $\overline{Z}^* \leq \overline{Z}$. Otherwise, there is no known necessary and sufficient conditions on h so that $\overline{Z}^* = \overline{Z}$ for all $\overline{X}$ in $[a, b]$.

There is nothing wrong in using α-cuts and interval arithmetic to evaluate fuzzy functions. Surely, it is user, and computer friendly. However, we should be aware that whenever we use α-cuts plus interval arithmetic to compute $\overline{Z} = H(\overline{X})$ we may be getting something larger than that obtained from the extension principle. The same results hold for functions of two or more independent variables.

Example 2.4.3.1

The example is the simple fuzzy expression

$$\overline{Z} = (1 - \overline{X})\, \overline{X}, \tag{2.33}$$

for $\overline{X}$ a triangular fuzzy number in $[0, 1]$. Let $\overline{X}[\alpha] = [x_1(\alpha), x_2(\alpha)]$. Using interval arithmetic we obtain

$$z_1(\alpha) = (1 - x_2(\alpha))x_1(\alpha), \tag{2.34}$$
$$z_2(\alpha) = (1 - x_1(\alpha))x_2(\alpha), \tag{2.35}$$

for $\overline{Z}[\alpha] = [z_1(\alpha), z_2(\alpha)]$, α in $[0, 1]$.

The extension principle extends the regular equation $z = (1 - x)x$, $0 \leq x \leq 1$, to fuzzy numbers as follows

$$\overline{Z}^*(z) = \sup_x \left\{ \overline{X}(x) | (1 - x)x = z, \ 0 \leq x \leq 1 \right\} . \tag{2.36}$$

Let $\overline{Z}^*[\alpha] = [z_1^*(\alpha), z_2^*(\alpha)]$. Then

$$z_1^*(\alpha) = \min\{(1 - x)x | x \in \overline{X}[\alpha]\}, \tag{2.37}$$
$$z_2^*(\alpha) = \max\{(1 - x)x | x \in \overline{X}[\alpha]\}, \tag{2.38}$$

for all $0 \leq \alpha \leq 1$. Now let $\overline{X} = (0/0.25/0.5)$, then $x_1(\alpha) = 0.25\alpha$ and $x_2(\alpha) = 0.50 - 0.25\alpha$. Equations (2.34) and (2.35) give $\overline{Z}[0.50] = [5/64, 21/64]$ but equations (2.37) and (2.38) produce $\overline{Z}^*[0.50] = [7/64, 15/64]$. Therefore, $\overline{Z}^* \neq \overline{Z}$. We do know that if each fuzzy number appears only once in the fuzzy expression, the two methods produce the same results ([8],[18]). However, if a fuzzy number is used more than once, as in equation (2.33), the two procedures can give different results.

2.5 Finding the Min/Max of a Fuzzy Number

In Chapters 13-17 we will want to determine the values of some decision variables $y = (x_1, ..., x_n)$ that will minimize (or maximize) a fuzzy function $\overline{E}(y)$. For each value of y we obtain a fuzzy number $\overline{E}(y)$. The following discussion will focus on $min\overline{E}(y)$ and at the end of the section we show what changes are needed for $max\overline{E}(y)$.

We can not minimize a fuzzy number so what we are going to do, which we have done before ([6],[9]-[13]), is first change $min\overline{E}(y)$ into a multiobjective problem and then translate the multiobjective problem into a single objective problem. This strategy is adopted from the finance literature where they had the problem of minimizing a random variable X whose values are constrained by a probability density function $g(x)$. They considered the multiobjective problem: (1) minimize the expected value of X; (2) minimize the variance of X; and (3) minimize the skewness of X to the right of the expected value. For our problem let: (1) $c(y)$ be the center of the core of $\overline{E}(y)$, the core of a fuzzy number is the interval where the membership function equals one, for each y; (2) $L(y)$ be the area under the graph of the membership function to the left of $c(y)$; and (3) $R(y)$ be the area under the graph of the membership function to the right of $c(y)$. See Figure 2.5. For $min\overline{E}(y)$ we substitute: (1) $min[c(y)]$; (2) $maxL(y)$, or maximize the possibility of obtaining values less than $c(y)$; and (3) $minR(y)$, or minimize the possibility of obtaining values greater then $c(y)$. So for $min\overline{E}(y)$ we have

$$V = (maxL(y), min[c(y)], minR(y)). \tag{2.39}$$

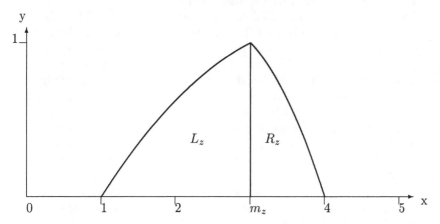

Figure 2.5: Computations for the Minimum of a Fuzzy Number

First let M be a sufficiently large positive number so that $maxL(y)$ is equivalent to $minL^*(y)$ where $L^*(y) = M - L(y)$. The multiobjective problem becomes

$$minV' = (minL^*(y), min[c(y)], minR(y)). \tag{2.40}$$

In a multiobjective optimization problem a solution is a value of the decision variable y that produces an undominated vector V'. Let $\mathcal{V}$ be the set of all vectors V' obtained for all possible values of the decision variable y. Vector $v_a = (v_{a1}, v_{a2}, v_{a3})$ dominates vector $v_b = (v_{b1}, v_{b2}, v_{b3})$, both in $\mathcal{V}$, if $v_{ai} \leq v_{bi}$, $1 \leq i \leq 3$, with one of the $\leq$ a strict inequality $<$. A vector $v \in \mathcal{V}$ is undominated if no $w \in \mathcal{V}$ dominates v. The set of undominated vectors in $\mathcal{V}$ is considered the general solution and the problem is to find values of the decision variables that produce undominated V'. The above definition of undominated was for a min problem, obvious changes need to be made for a max problem.

One way to explore the undominated set is to change the multiobjective problem into a single objective. The single objective problem is

$$min(\lambda_1 [M - L(y)] + \lambda_2 c(y) + \lambda_3 R(y)), \qquad (2.41)$$

where $\lambda_i > 0$, $1 \leq i \leq 3$, $\lambda_1 + \lambda_2 + \lambda_3 = 1$. You will get different undominated solutions by choosing different values of $\lambda_i > 0$, $\lambda_1 + \lambda_2 + \lambda_3 = 1$. It is known that solutions to this problem are undominated, but for some problems it will be unable to generate all undominated solutions [16]. The decision maker is to choose the values of the weights λ_i for the three minimization goals. Usually one picks different values for the λ_i to explore the solution set and then lets the decision maker choose an optimal y^* from this set of solutions. It is sometimes best to present management with a number of optimal solutions, instead of only one optimal solution. Managers are decision makes, and when they have multiple solutions to pick from, they can decide on one of them weighing the various alternatives associated with each.

This is how we propose to handle the problem of $min\overline{E}(y)$ in Chapters 13-17. Numerical solutions to this optimization problem can be difficult, depending on the variables and the constraints. In the past we have employed an evolutionary algorithm to generate good approximate solutions. See ([4],[5],[9]) for a general description of our evolutionary algorithm and other applications to solving fuzzy optimization problems. However, the fuzzy optimization problems in Chapters 13-17 are all discrete fuzzy optimization problems. By discrete we mean that there are only a finite number of possible values for the decision variable y. These fuzzy optimization problems are easier to solve and we will not need to employ a evolutionary algorithm in their solution.

Now consider $max\overline{E}(y)$. Then we would want to $minL(y)$, $max[c(y)]$ and $maxR(y)$. The objective function is equation (2.41) after changing min to max.

2.6 Ordering/Ranking Fuzzy Numbers

Given a finite set of fuzzy numbers $\overline{A}_1, ..., \overline{A}_n$ in Chapters 13-17, we want to order/rank them from smallest to largest. Each $\overline{A}_i$ corresponds to a decision variable a_i, $1 \leq i \leq n$, and in a *max (min)* problem the largest (smallest) $\overline{A}_i$ gives the optimal choice for the decision variables. For a finite set of real numbers there is no problem in ordering them from smallest to largest. However, in the fuzzy case there is no universally accepted way to do this. There are probably more than 40 methods proposed in the literature of defining $\overline{M} \leq \overline{N}$, for two fuzzy numbers $\overline{M}$ and $\overline{N}$. Here the symbol $\leq$ means "less than or equal" and not "a fuzzy subset of". A few key references on this topic are ([1],[14],[15],[20],[21]), where the interested reader can look up many of these methods and see their comparisons.

Here we will present only one procedure for ordering fuzzy numbers that we have used before ([2],[3]). But note that different definitions of $\leq$ between fuzzy numbers can give different ordering. We first define $\leq$ between two fuzzy numbers $\overline{M}$ and $\overline{N}$. Define

$$v(\overline{M} \leq \overline{N}) = max\{min(\overline{M}(x), \overline{N}(y))|x \leq y\}, \qquad (2.42)$$

which measures how much $\overline{M}$ is less than or equal to $\overline{N}$. We write $\overline{N} < \overline{M}$ if $v(\overline{N} \leq \overline{M}) = 1$ but $v(\overline{M} \leq \overline{N}) < \eta$, where η is some fixed fraction in $(0, 1]$. In this book we will use $\eta = 0.8$. Then $\overline{N} < \overline{M}$ if $v(\overline{N} \leq \overline{M}) = 1$ and $v(\overline{M} \leq \overline{N}) < 0.8$. We then define $\overline{M} \approx \overline{N}$ when both $\overline{N} < \overline{M}$ and $\overline{M} < \overline{N}$ are false. $\overline{M} \leq \overline{N}$ means $\overline{M} < \overline{N}$ or $\overline{M} \approx \overline{N}$. Now this $\approx$ may not be transitive. If $\overline{N} \approx \overline{M}$ and $\overline{M} \approx \overline{O}$ implies that $\overline{N} \approx \overline{O}$, then $\approx$ is transitive. However, it can happen that $\overline{N} \approx \overline{M}$ and $\overline{M} \approx \overline{O}$ but $\overline{N} < \overline{O}$ because $\overline{M}$ lies a little to the right of $\overline{N}$ and $\overline{O}$ lies a little to the right of $\overline{M}$ but $\overline{O}$ lies sufficiently far to the right of $\overline{N}$ that we obtain $\overline{N} < \overline{O}$. But this ordering is still useful in partitioning the set of fuzzy numbers up into sets $H_1, ..., H_K$ where ([2],[3]): (1) given any $\overline{M}$ and $\overline{N}$ in H_k, $1 \leq k \leq K$, then $\overline{M} \approx \overline{N}$; and (2) given $\overline{N} \in H_i$ and $\overline{M} \in H_j$, with $i < j$, then $\overline{N} \leq \overline{M}$. Then the highest ranked fuzzy numbers lie in H_K, the second highest ranked fuzzy numbers are in H_{K-1}, etc. This result is easily seen if you graph all the fuzzy numbers on the same axis then those in H_K will be clustered together farthest to the right, proceeding from the H_K cluster to the left the next cluster will be those in H_{K-1}, etc. Then in a *max (min)* decision problem the optimal values of the decision variables correspond to those fuzzy sets in H_K (H_1). If you require a unique decision, then you will need to decide between those fuzzy numbers in the highest (lowest) ranked set, possibly using the model in the previous section.

Let $\overline{N} \in H_i$ and $\overline{M} \in H_j$ with $i < j$. Then $\overline{N} \leq \overline{M}$, which means that $\overline{N} < \overline{M}$ or $\overline{N} \approx \overline{M}$. It may happen that, and this does occur in Chapters 13, 14 and 16, $\overline{N} \approx \overline{M}$. But then we can always find a $\overline{O} \in H_j$ so that $\overline{N} < \overline{O}$. What this means is that we may get $\overline{N}$ in H_i approximately equal to an $\overline{M}$ in H_j with $i < j$, but this $\overline{N}$ will not be approximately equal to all the fuzzy numbers in H_j.

There is an easy way to determine if $\overline{M} < \overline{N}$, or $\overline{M} \approx \overline{N}$, for many fuzzy numbers. First, it is easy to see that if the core of $\overline{N}$ lies completely to the

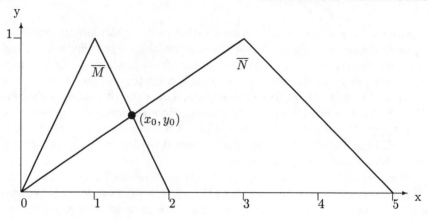

Figure 2.6: Determining $v(\overline{N} \leq \overline{M})$

right of the core of $\overline{M}$, then $v(\overline{M} \leq \overline{N}) = 1$. Also, if the core of $\overline{M}$ and the core of $\overline{N}$ overlap, then $\overline{M} \approx \overline{N}$. Now assume that the core of $\overline{N}$ lies to the right of the core of $\overline{M}$, as shown in Figure 2.6 for triangular fuzzy numbers, and we wish to compute $v(\overline{N} \leq \overline{M})$. The value of this expression is simply y_0 in Figure 2.6. In general, for triangular (shaped), and trapezoidal (shaped), fuzzy numbers $v(\overline{N} \leq \overline{M})$ is the height of their intersection when the core of $\overline{N}$ lies to the right of the core of $\overline{M}$. Locate η, for example $\eta = 0.8$ in this book, on the vertical axis and then draw a horizontal line through η. If in Figure 2.6 y_0 lies below the horizontal line, then $\overline{M} < \overline{N}$. If y_0 lies on, or above, the horizontal line, then $\overline{M} \approx \overline{N}$.

2.7 References

1. G.Bortolon and R.Degani: A Review of Some Methods for Ranking Fuzzy Subsets, Fuzzy Sets and Systems, 15(1985), pp. 1-19.

2. J.J.Buckley: Ranking Alternatives Using Fuzzy Numbers, Fuzzy Sets and Systems, 15(1985), pp.21-31.

3. J.J.Buckley: Fuzzy Hierarchical Analysis, Fuzzy Sets and Systems, 17(1985), pp. 233-247.

4. J.J.Buckley and E.Eslami: Introduction to Fuzzy Logic and Fuzzy Sets, Physica-Verlag, Heidelberg, Germany, 2002.

5. J.J.Buckley and T.Feuring: Fuzzy and Neural: Interactions and Applications, Physica-Verlag, Heidelberg, Germany, 1999.

6. J.J.Buckley and T.Feuring: Evolutionary Algorithm Solutions to Fuzzy Problems: Fuzzy Linear Programming, Fuzzy Sets and Systems, 109(2000), pp. 35-53.

7. J.J. Buckley and Y. Hayashi: Can Neural Nets be Universal Approximators for Fuzzy Functions?, Fuzzy Sets and Systems, 101 (1999), pp. 323-330.

8. J.J. Buckley and Y. Qu: On Using α-cuts to Evaluate Fuzzy Equations, Fuzzy Sets and Systems, 38(1990), pp. 309-312.

9. J.J.Buckley, E.Eslami and T.Feuring: Fuzzy Mathematics in Economics and Engineering, Physica-Verlag, Heidelberg, Germany, 2002.

10. J.J.Buckley, T.Feuring and Y.Hayashi: Solving Fuzzy Problems in Operations Research, J. Advanced Computational Intelligence, 3(1999), pp. 171-176.

11. J.J.Buckley, T.Feuring and Y.Hayashi: MultiObjective Fully Fuzzified Linear Programming, Int. J. Uncertainty, Fuzziness and Knowledge Based Systems, 9(2001), pp. 605-622.

12. J.J.Buckley, T.Feuring and Y.Hayashi: Fuzzy Queuing Theory Revisited", Int. J. Uncertainty, Fuzziness and Knowledge Based Systems, 9(2001), pp. 527-538.

13. J.J.Buckley, T.Feuring and Y.Hayashi: Solving Fuzzy Problems in Operations Research: Inventory Control, Soft Computing, 7(2002), pp. 121-129.

14. P.T.Chang and E.S.Lee: Fuzzy Arithmetic and Comparison of Fuzzy Numbers, in: M.Delgado, J.Kacprzyk, J.L.Verdegay and M.A.Vila (eds.), Fuzzy Optimization: Recent Advances, Physica-Verlag, Heidelberg, Germany, 1994, pp. 69-81.

15. D.Dubois, E.Kerre, R.Mesiar and H.Prade: Fuzzy Interval Analysis, in: D.Dubois and H.Prade (eds.), Fundamentals of Fuzzy Sets, The Handbook of Fuzzy Sets, Kluwer Acad. Publ., 2000, pp. 483-581.

16. A.M.Geoffrion: Proper Efficiency and the Theory of Vector Maximization, J. Math. Analysis and Appl., 22(1968), pp. 618-630.

17. G.J. Klir and B. Yuan: Fuzzy Sets and Fuzzy Logic: Theory and Applications, Prentice Hall, Upper Saddle River, N.J., 1995.

18. R.E. Moore: Methods and Applications of Interval Analysis, SIAM Studies in Applied Mathematics, Philadelphia, 1979.

19. A. Neumaier: Interval Methods for Systems of Equations, Cambridge University Press, Cambridge, U.K., 1990.

20. X.Wang and E.E.Kerre: Reasonable Properties for the Ordering of Fuzzy Quantities (I), Fuzzy Sets and Systems, 118(2001), pp. 375-385.

21. X.Wang and E.E.Kerre: Reasonable Properties for the Ordering of Fuzzy Quantities (II), Fuzzy Sets and Systems, 118(2001), pp. 387-405.

Chapter 3

Fuzzy Probabilities/Arrival Rates

3.1 Introduction

The first thing to do is explain how we will get fuzzy probabilities, which will be fuzzy numbers, from a set of confidence intervals. This is done in the next section. Next we discuss how we can obtain fuzzy numbers for arrival rates and for service rates in Section 3.3. Then we discuss "restricted fuzzy arithmetic" in Section 3.5. Throughout this book whenever we wish to find the α-cut of a fuzzy probability, or a certain fuzzy number, we usually need to solve an optimization problem. We discuss this computation problem in more detail in Section 3.6. The last section is about how we constructed figures of fuzzy probabilities, or certain fuzzy numbers, using Maple [10], or $LaTeX2_\epsilon$.

3.2 Fuzzy Probabilities from Confidence Intervals

We will measure changes in our system at time intervals δ. This time interval may be one second, or 0.1 seconds, etc. The use of δ is explained in more detail in the next chapter where we introduce fuzzy finite Markov chains. We first need to gather data about the system, like the probability that i customers arrive during time interval δ. We assume that the system is in its usual operating mode and not in a "burstiness" period (Chapter 15) nor does it have some customers who take a long time in a server (Chapter 16). The book [9] describes both "burstiness" and "long tailed distributions". However, we can obtain fuzzy probabilities in these cases just as described below. Suppose we observe the system during N time periods and find that there have been

n_i times that i customers have arrived for service, $i = 0, 1, 2, 3....$ We would expect, from practical considerations, that there is some positive integer L so that $n_i = 0$ for $i > L$. Let $p(i)$ be the probability that i customers arrive during time period δ, $i = 0, 1, 2, 3, ..., L$. Then a point estimate of $p(i)$ is simply n_i/N. However, to show our uncertainty in this estimate we may also compute a confidence interval for $p(i)$.

We propose to find the $(1 - \beta)100\%$ confidence interval for $p(i)$, for all $0.01 \leq \beta < 1$. Starting at 0.01 is arbitrary and you could begin at 0.001, or 0.005, etc. Denote these confidence intervals as

$$[p(i)_1(\beta), p(i)_2(\beta)], \tag{3.1}$$

for $0.01 \leq \beta < 1$. Add to this the interval $[n_i/N, n_i/N]$ for the 0% confidence interval for $p(i)$. Then we have $(1 - \beta)100\%$ confidence interval for $p(i)$ for $0.01 \leq \beta \leq 1$.

Now place these confidence intervals, one on top of the other, to produce a triangular shaped fuzzy number $\overline{p}(i)$ whose α-cuts are the confidence intervals. We have

$$\overline{p}(i)[\alpha] = [p(i)_1(\alpha), p(i)_2(\alpha)], \tag{3.2}$$

for $0.01 \leq \alpha \leq 1$. All that is needed is to finish the "bottom" of $\overline{p}(i)$ to make it a complete fuzzy number. We will simply drop the graph of $\overline{p}(i)$ straight down to complete its α-cuts so

$$\overline{p}(i)[\alpha] = [p(i)_1(0.01), p(i)_2(0.01)], \tag{3.3}$$

for $0 \leq \alpha < 0.01$. In this way we are using more information in $\overline{p}(i)$ than just a point estimate, or just a single interval estimate. Notice that $\overline{p}(i)[0]$ is the 99% confidence interval for $p(i)$.

The following example shows that the $\overline{p}(i)$ will be a triangular shaped fuzzy number. However, for simplicity, throughout the rest of this book we will always use triangular fuzzy numbers for the fuzzy values of uncertain probabilities.

Example 3.2.1

Suppose $N = 500$ and $n_1 = 100$. Using the normal approximation to the binomial an approximate $(1 - \beta)100\%$ confidence interval for $p(1)$ is

$$[0.2 - z_{\beta/2}\sqrt{\frac{0.2(1 - 0.2)}{500}}, 0.2 + z_{\beta/2}\sqrt{\frac{0.2(1 - 0.2)}{500}}], \tag{3.4}$$

where $z_{\beta/2}$ is defined as

$$\int_{-\infty}^{z_{\beta/2}} N(0, 1)dx = 1 - \beta/2, \tag{3.5}$$

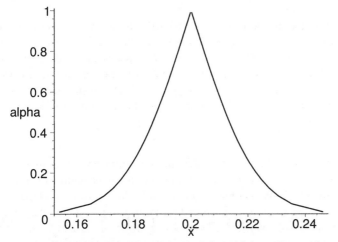

Figure 3.1: Fuzzy Probability $\bar{p}(1)$ in Example 3.2.1

and $N(0,1)$ denotes the normal density with mean zero and unit variance. We evaluated equations (3.4) and (3.5) using Maple [10] and then the graph of $\bar{p}(1)$ is shown in Figure 3.1 , without dropping the graph straight down to the x-axis at the end points. In this way we obtain triangular shaped fuzzy numbers for $\bar{p}(i)$, $i = 0, 1, 2, 3, ..., L$. However, in further calculations we will be using triangular fuzzy numbers for the $\bar{p}(i)$.

3.3 Fuzzy Arrival/Service Rates

In this section we concentrate on deriving fuzzy numbers for the arrival rate, and the service rate, in a queuing system. We also assume, as in the previous section, that we are in its usual operating mode and not in a "burstiness" period (Chapter 15) nor does it have some customers who take a long time in a server (Chapter 16). But, we may find the fuzzy probabilities for theses cases in the same manner as presented below. We consider the fuzzy arrival rate first.

3.3.1   Fuzzy Arrival Rate

We assume that we have Poisson arrivals [13] which means that there is a positive constant λ so that the probability of k arrivals per unit time is

$$\lambda^k \exp(-\lambda)/k!, \tag{3.6}$$

the Poisson probability function. We need to estimate λ, the arrival rate, so we take a random sample $X_1, ..., X_m$ of size m. In the random sample X_i is the number of arrivals per unit time, in the ith observation. Let S be the sum of the X_i and let $\overline{X}$ be S/m. Here, $\overline{X}$ is not a fuzzy set but the mean.

Now S is Poisson with parameter $m\lambda$ ([4], p. 298). Assuming that $m\lambda$ is sufficiently large (say, at least 30), we may use the normal approximation ([4], p. 317), so the statistic

$$W = \frac{S - m\lambda}{\sqrt{m\lambda}}, \tag{3.7}$$

is approximately a standard normal. Then

$$P[-z_{\beta/2} < W < z_{\beta/2}] = 1 - \beta, \tag{3.8}$$

where the $z_{\beta/2}$ was defined in equation (3.5). Now divide numerator and denominator of W by m and we get

$$P[-z_{\beta/2} < Z < z_{\beta/2}] = 1 - \beta, \tag{3.9}$$

where

$$Z = \frac{\overline{X} - \lambda}{\sqrt{\lambda/m}}. \tag{3.10}$$

From these last two equations we may derive an approximate $(1 - \beta)100\%$ confidence interval for λ. Let us call this confidence interval $[l(\beta), r(\beta)]$.

We now show how to compute $l(\beta)$ and $r(\beta)$. Let

$$f(\lambda) = \sqrt{m}(\overline{X} - \lambda)/\sqrt{\lambda}. \tag{3.11}$$

Now $f(\lambda)$ has the following properties: (1) it is strictly decreasing for $\lambda > 0$; (2) it is zero for $\lambda > 0$ only at $\overline{X} = \lambda$; (3) the limit of f, as λ goes to ∞ is $-\infty$; and (4) the limit of f as λ approaches zero from the right is ∞. Hence, (1) the equation $z_{\beta/2} = f(\lambda)$ has a unique solution $\lambda = l(\beta)$; and (2) the equation $-z_{\beta/2} = f(\lambda)$ also has a unique solution $\lambda = r(\beta)$.

We may find these unique solutions. Let

$$V = \sqrt{z_{\beta/2}^2/m + 4\overline{X}}, \tag{3.12}$$

$$z_1 = [-\frac{z_{\beta/2}}{\sqrt{m}} + V]/2, \tag{3.13}$$

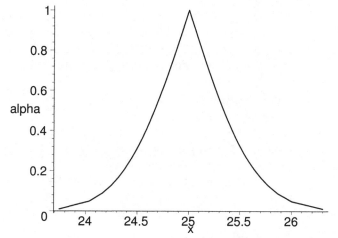

Figure 3.2: Fuzzy Arrival Rate $\overline{\lambda}$ in Example 3.3.1.1

and

$$z_2 = [\frac{z_{\beta/2}}{\sqrt{m}} + V]/2. \tag{3.14}$$

Then $l(\beta) = z_1^2$ and $r(\beta) = z_2^2$.

We now substitute α for β to get the α-cuts of fuzzy number $\overline{\lambda}$. Add the point estimate, when $\alpha = 1$, $\overline{X}$, for the 0% confidence interval. Now as α goes from 0.01 (99% confidence interval) to one (0% confidence interval) we get the fuzzy number for λ. As before, we drop the graph straight down at the ends to obtain a complete fuzzy number.

Example 3.3.1.1

Suppose $m = 100$ and we obtained $\overline{X} = 25$. We evaluated equations (3.12) through (3.14) using Maple [10] and then the graph of $\overline{\lambda}$ is shown in Figure 3.2, without dropping the graph straight down to the x−axis at the end points. However, starting in Chapter 11 we will use a triangular fuzzy number for $\overline{\lambda}$.

3.3.2   Fuzzy Service Rate

Let μ be the average (expected) service rate, in the number of service completions per unit time, for a busy server. Then $1/\mu$ is the average (expected) service time. The probability density of the time interval between successive service completions is ([13], Chapter 15)

$$(1/\mu)\exp(-t/\mu), \tag{3.15}$$

for $t > 0$, the exponential probability density function. Let $X_1, ..., X_n$ be a random sample from this exponential density function. Then the maximum likelihood estimator for μ is $\overline{X}$ ([4],p.344), the mean of the random sample (not a fuzzy set). We know that the probability density for $\overline{X}$ is the gamma ([4],p.297) with mean μ and variance μ^2/n ([4],p.351). If n is sufficiently large we may use the normal approximation to determine approximate confidence intervals for μ. Let

$$Z = (\sqrt{n}[\overline{X} - \mu])/\mu, \tag{3.16}$$

which is approximately normally distributed with zero mean and unit variance, provided n is sufficiently large. See Figure 6.4-2 in [4] for $n = 100$ which shows the approximation is quite good if $n = 100$. The graph in Figure 6.4-2 in [4] is for the chi-square distribution which is a special case of the gamma distribution. So we now assume that $n \geq 100$ and use the normal approximation to the gamma.

An approximate $(1 - \beta)100\%$ confidence interval for μ is obtained from

$$P[-z_{\beta/2} < Z < z_{\beta/2}] = 1 - \beta, \tag{3.17}$$

where β was defined in equation (3.5). After solving for μ we get

$$P[L(\beta) < \mu < R(\beta)] = 1 - \beta, \tag{3.18}$$

where

$$L(\beta) = [\sqrt{n}\,\overline{X}]/[z_{\beta/2} + \sqrt{n}], \tag{3.19}$$

and

$$R(\beta) = [\sqrt{n}\,\overline{X}]/[\sqrt{n} - z_{\beta/2}]. \tag{3.20}$$

An approximate $(1 - \beta)100\%$ confidence interval for μ is

$$[\frac{\sqrt{n}\,\overline{X}}{z_{\beta/2} + \sqrt{n}}, \frac{\sqrt{n}\,\overline{X}}{\sqrt{n} - z_{\beta/2}}]. \tag{3.21}$$

Example 3.3.2.1

If $n = 400$ and $\overline{X} = 1.5$, then we get

$$[\frac{30}{z_{\beta/2} + 20}, \frac{30}{20 - z_{\beta/2}}], \tag{3.22}$$

for a $(1 - \beta)100\%$ confidence interval for the service rate μ. Now we can put these confidence intervals together, one on top of another, to obtain a fuzzy number $\overline{\mu}$ for the service rate. We evaluated equation (3.22) using Maple [10] for $0.01 \leq \beta \leq 1$ and the graph of the fuzzy service rate, without dropping the graph straight down to the x-axis at the end points, is in Figure 3.3. For simplicity we use triangular fuzzy numbers for $\overline{\mu}$ in the rest of the book.

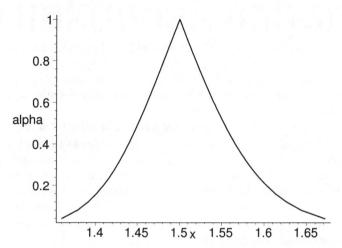

Figure 3.3: Fuzzy Service Rate $\overline{\mu}$ in Example 3.3.2.1

3.4 Fuzzy Numbers from Expert Opinion

Some of the fuzzy probabilities and fuzzy constants in our models may have to be estimated by experts. That is, we have no statistical data to generate the fuzzy numbers from a set of confidence intervals as discussed in the previous sections. So let us briefly see how this may be accomplished. First assume we have only one expert and he/she is to estimate the value of some probability p. We can solicit this estimate from the expert as is done in estimating job times in project scheduling ([13], Chapter 13). Let a = the "pessimistic" value of p, or the smallest possible value, let c = be the "optimistic" value of p, or the highest possible value, and let b = the most likely value of p. We then ask the expert to give values for a, b, c and we construct the triangular fuzzy number $\overline{p} = (a/b/c)$ for p. If we have a group of N experts all to estimate the value of p we solicit the a_i, b_i and c_i, $1 \leq i \leq N$, from them. Let a be the average of the a_i, b is the mean of the b_i and c is the average of the c_i. The simplest thing to do is to use $(a/b/c)$ for p. The same method can be used to estimate other constants presented in the rest of the book.

3.5 Restricted Fuzzy Arithmetic

Restricted fuzzy arithmetic was first proposed in the papers ([5]-[7],[11],[12], see also [8]). We will first discuss three methods that you may obtain crisp probabilities. Next we explain what we mean by restricted fuzzy arithmetic for a finite, discrete, fuzzy probability distribution ([1],[2]). Then we will extend it to how it will be used in this book. Restricted fuzzy arithmetic is not used in Chapters 11-12 where we have fuzzy arrival/service rates.

3.5.1   Probabilities

Let $X = \{x_1, ..., x_n\}$ be a finite set and let P be a probability function defined on all subsets of X with $P(\{x_i\}) = a_i$, $1 \leq i \leq n$, $0 < a_i < 1$, all i, and $\sum_{i=1}^{n} a_i = 1$. Starting in Chapter 4 we will substitute a fuzzy number $\overline{a}_i$ for a_i, for some i, to obtain a discrete (finite) fuzzy probability distribution. Where do these fuzzy numbers come from?

In some problems, because of the way the problem is stated, the values of all the a_i are crisp and known. For example, consider tossing a fair coin and $a_1 = $ the probability of getting a "head" and $a_2 = $ is the probability of obtaining a "tail". Since we assumed it to be a fair coin we must have $a_1 = a_2 = 0.5$. In this case we would not substitute a fuzzy number for a_1 or a_2 because they are crisp numbers 0.5. But in many other problems the a_i are not known exactly and they are either estimated from a random sample or they are obtained from "expert opinion".

Suppose we have the results of a random sample to estimate the value of a_1. We would construct a set of confidence intervals for a_1 and then put these together to get the fuzzy number $\overline{a}_1$ for a_1. This method of building a fuzzy number from confidence intervals was discussed in Section 3.2.

Assume that we do not know the values of the a_i and we do not have any data to estimate their values. Then we may obtain numbers for the a_i from some group of experts. This group could consist of only one expert. This case includes subjective, or "personal", probabilities and it is used to estimate certain probabilities in this book and certain parameters (like revenue, costs,...) in Chapters 13-17. We discussed this case in Section 3.4 above.

So, when we have to estimate probabilities/parameters from data/expert opinion, we will use fuzzy numbers for these uncertain items.

3.5.2   Restricted Arithmetic: General

Let $X = \{x_1, ..., x_n\}$ be a finite set and let P be a probability function defined on all subsets of X with $P(\{x_i\}) = a_i$, $1 \leq i \leq n$, $0 < a_i < 1$ all i and $\sum_{i=1}^{n} a_i = 1$. X together with P is a discrete (finite) probability distribution. In practice all the a_i values must be known exactly. Many times these values are estimated, or they are provided by experts. We now assume that some of these a_i values are uncertain and we will model this uncertainty using fuzzy numbers. Not all the a_i need to be uncertain, some may be known exactly and are given as a crisp (real) number. If an a_i is crisp, then we will still write it as a fuzzy number even though this fuzzy number is crisp.

Due to the uncertainty in the a_i values we substitute $\overline{a}_i$, a fuzzy number, for each a_i and assume that $0 < \overline{a}_i < 1$ all i. If some a_i is known precisely, then this $\bar{a}_i = a_i$ but we still write a_i as $\bar{a}_i$. Then X together with the $\bar{a}_i$ values is a discrete (finite) fuzzy probability distribution. We write $\bar{P}$ for fuzzy P and we have $\bar{P}(\{x_i\}) = \bar{a}_i$, $1 \leq i \leq n$, $0 < \bar{a}_i < 1$.

The uncertainty is in some of the a_i values but we know that we have a discrete probability distribution. So we now put the following restriction on the $\bar{a}_i$ values: there are $a_i \in \bar{a}_i[1]$ so that $\sum_{i=1}^n a_i = 1$. That is, we can choose a_i in $\bar{a}_i[\alpha]$, all α, so that we get a discrete probability distribution.

Let A and B be (crisp) subsets of X. We know how to compute $P(A)$ and $P(B)$ so let us find $\bar{P}(A)$ and $\bar{P}(B)$, the fuzzy probabilities of A and B, respectively. To do this we introduce restricted fuzzy arithmetic. There may be uncertainty in some of the a_i values, but there is no uncertainty in the fact that we have a discrete probability distribution. That is, whatever the a_i values in $\bar{a}_i[\alpha]$ we must have $a_1 + ... + a_n = 1$. This is the basis of our restricted fuzzy arithmetic. Suppose $A = \{x_1, ..., x_k\}$, $1 \leq k < n$, then define (this is an α-cut of the fuzzy probability)

$$\bar{P}(A)[\alpha] = \{\sum_{i=1}^k a_i | \quad \mathbf{S} \quad \}, \tag{3.23}$$

for $0 \leq \alpha \leq 1$, where $\mathbf{S}$ stands for the statement "$a_i \in \bar{a}_i[\alpha]$, $1 \leq i \leq n$, $\sum_{i=1}^n a_i = 1$ ". This is our restricted fuzzy arithmetic. Notice that we first choose a complete discrete probability distribution from the α-cuts before we compute a probability in equation (3.23). Notice also that $\bar{P}(A)[\alpha]$ is not the sum of the intervals $\bar{a}_i[\alpha]$, $1 \leq i \leq k$, using interval arithmetic. The α-cuts defined in equation (3.23) are then put together to obtain a fuzzy number for the fuzzy probability. $\bar{P}(A)[\alpha]$ will be an interval, and we will use the right side of equation (3.23) to compute the end points of this interval.

3.5.3 Restricted Fuzzy Arithmetic: Book

Now let us look at a typical restricted arithmetic calculation in Chapters 6-10. We want to compute α-cuts of a fuzzy probability $\bar{p}_{ij}$ in a 5×5 fuzzy matrix. We have two fuzzy, discrete, probability distributions: (1) $\bar{p}_i$, $1 \leq i \leq 7$; and (2) $\bar{q}_i$, $i = 1, 2$. The equation for some $\bar{p}_{ij}$ is

$$\bar{p}_{ij} = \bar{p}_1 \bar{q}_1 + \bar{p}_2 \bar{q}_2. \tag{3.24}$$

We do not substitute the fuzzy numbers for $\bar{p}_i$ and $\bar{q}_i$ into equation (3.24) to find $\bar{p}_{ij}$ because we may get a fuzzy number not entirely in $[0, 1]$. We use restricted fuzzy arithmetic to get α-cuts of $\bar{p}_{ij}$

$$\bar{p}_{ij}[\alpha] = \{p_1 q_1 + p_2 q_2 | \quad \mathbf{S} \quad \}, \tag{3.25}$$

where $\mathbf{S}$ is the statement " $p_i \in \bar{p}_i[\alpha]$, $1 \leq i \leq 7$, $p_1 + ... + p_7 = 1$, $q_i \in \bar{q}_i[\alpha]$, $i = 1, 2$, $q_1 + q_2 = 1$". $\bar{p}_{ij}[\alpha]$ is an interval, say $[\tau_1(\alpha), \tau_2(\alpha)]$, where

$$\tau_1(\alpha) = min\{p_1 q_1 + p_2 q_2 | \quad \mathbf{S} \quad \}, \tag{3.26}$$

and

$$\tau_2(\alpha) = max\{p_1q_1 + p_2q_2 | \quad \mathbf{S} \quad \}. \tag{3.27}$$

Methods for finding the min (max) in equation (3.26) ((3.27)) are discussed in the next section.

Restricted fuzzy arithmetic, extended to restricted fuzzy matrix multiplication, is used in the next chapter to get the fuzzy steady state probabilities. All the fuzzy computations needed to get to the fuzzy numbers for system performance are described in the Chapters 6-10 and in Chapter 19.

3.6 Computations

Throughout this book whenever we wish to find the α-cut of a fuzzy probability, or some other fuzzy number, we usually need to solve an optimization problem. This was seen in the previous section in computing $\overline{p}_{ij}[\alpha]$. Let us look now in more detail at three situations that will arise later on in the book. Other computational problems will be discussed at the point were they appear. In particular, those computations related to using fuzzy arrival/service rates will be discussed in Chapters 11 and 12.

3.6.1 First Problem

This is the problem in equations (3.25)-(3.27) in the previous Subsection 3.5.3. This is were we want to find α-cuts of $\overline{p}_{ij}$ were it equals $\overline{p}_1\overline{q}_1 + \overline{p}_2\overline{q}_2$. If $\overline{p}_{ij}[\alpha] = [\tau_1(\alpha), \tau_2(\alpha)]$ and $\overline{p}_i[\alpha] = [p_{i1}(\alpha), p_{i1}(\alpha)]$, $i \leq i \leq 7$, and $\overline{q}_i[\alpha] = [q_{i1}(\alpha), q_{i2}(\alpha)]$, $i = 1, 2$, then the optimization problems are

$$\tau_1(\alpha) = min\{p_1q_1 + p_2q_2\}, \tag{3.28}$$

subject to

$$p_{i1}(\alpha) \leq p_i \leq p_{i2}(\alpha), 1 \leq i \leq 7, p_1 + \ldots + p_7 = 1, \tag{3.29}$$

and

$$q_{i1}(\alpha) \leq q_i \leq q_{i2}(\alpha), i = 1, 2, q_1 + q_2 = 1, \tag{3.30}$$

and

$$\tau_2(\alpha) = max\{p_1q_1 + p_2q_2\}, \tag{3.31}$$

subject to the same constraints, for all α in $[0, 1]$. The min problem gives the left point of the interval and the max problem produces the right end point.

Let $f = p_1q_1 + p_2q_2$. Clearly, f is an increasing function of p_1, p_2, q_1, q_2. So our first choice is $p_1 = p_{11}(\alpha)$, $p_2 = p_{21}(\alpha)$, $q_1 = q_{11}(\alpha)$ and $q_2 = q_{21}(\alpha)$ to get $\tau_1(\alpha)$ and choose the right end points of the intervals for $\tau_2(\alpha)$. However, we have not checked to see if the choices are "feasible".

The choice of $p_1 = p_{11}(\alpha)$, $p_2 = p_{21}(\alpha)$ is feasible if we can get $p_i \in \overline{p}_i(\alpha)$, $3 \leq i \leq 7$ so that $p_1 + \ldots + p_7 = 1$. The choice of $q_1 = q_{11}(\alpha)$ and $q_2 = q_{21}(\alpha)$

is feasible when their sum is one. If both choices are feasible, then we have the answer for the min problem $\tau_1(\alpha)$. Otherwise, we need to use some optimization algorithm to obtain the left end point.

We also need to check our choices for the right end point $\tau_2(\alpha)$ to see if they are feasible. Otherwise, use the numerical algorithm.

Now assume that the choices for q_i were feasible but the choices for the p_i were not feasible. Then the optimization problem, for $\tau_1(\alpha)$, becomes

$$min\{p_1q_{11}(\alpha) + p_2q_{21}(\alpha)\}, \qquad (3.32)$$

subject to only the constraints on the p_i given in equation (3.29). Notice that the $q_{i1}(\alpha)$, $i = 1, 2$, are constants , not variables, in the new optimization problem. But this is a linear programming problem which we can solve by calling the "simplex" module in Maple [10]. Similar remarks for the max problem. This reduction to a linear programming problem occurs a number of times in the book. For these optimization problems we must always check to see if our choices are feasible.

Occasionally we end up with a non-linear optimization problem. Suppose the choices for the p_i and for the q_i are both not feasible. Then we have the non-linear optimization problems in equations (3.28) and (3.31) to solve subject to the constraints in equations (3.29) and (3.30). Here we used the Premium Solver Platform V5.0 from Frontline Systems [3]. This software in an add-on to Excel.

3.6.2 Second Problem

Here we have a fuzzy, discrete, probability distribution $\overline{p}_i$, $0 \leq i \leq M$, over $X = \{0, 1, 2, ..., M\}$. We want to find the mean of this fuzzy probability distribution. The mean $\overline{\mu}$ is

$$\overline{\mu} = \sum_{k=0}^{M} k\overline{p}_k, \qquad (3.33)$$

which is evaluated by α-cuts and restricted fuzzy arithmetic. So

$$\overline{\mu}[\alpha] = \{\sum_{k=0}^{M} kp_k | \ \mathbf{S} \ \}, \qquad (3.34)$$

where $\mathbf{S}$ is "$p_i \in \overline{p}_i[\alpha]$, $0 \leq i \leq M$, and $p_0 + ... + p_M = 1$". Let $\overline{\mu}[\alpha] = [\mu_1(\alpha), \mu_2(\alpha)]$. The optimization problems are

$$\mu_1(\alpha) = min \sum_{k=0}^{M} kp_k, \qquad (3.35)$$

subject to the constraints in statement **S**, and

$$\mu_2(\alpha) = max \sum_{k=0}^{M} k p_k, \tag{3.36}$$

subject to the same constraints.

Both optimization problems are linear programming problems and hence can be solved using Maple.

3.6.3 Another Fuzzy Computation

Next we need to estimate the probability that a customer leaves a busy server during time interval δ. The servers will be parallel, independent and identical so we only need to gather data on one of the servers. Suppose a customer is in the server at the start of the time period. Let p be the probability that the customer finishes and leaves the server during the time period δ. We assume the servers have the Markov property, or they are "memoryless", which means that p does not depend on a customer having been in this server during the previous time period. Let N be the number of time periods that we have observed this server, which was busy at the start of the period, and let n be the number of times a customer finished and left the server. Then a point estimate of p is n/N. But, as was done in Section 3.2, we build a triangular shaped fuzzy number $\bar{p}$ for p from the confidence intervals. We also just use a triangular fuzzy number for $\bar{p}$ in future computations.

All we need for Chapters 4-10 are the fuzzy numbers for $\bar{p}(i)$, $0 \leq i \leq L$ and $\bar{p}$. However, before we go on we need to discuss how we will handle multiple servers. Let there be c parallel, independent and identical servers. Let $q(i|s)$ be the crisp probability that i customers leave the servers during a time interval δ, given that s servers were busy at the start of the time interval, $i = 0, 1, 2, ..., s$ and $s = 1, 2, 3, ..., c$. We have that $\sum_{i=0}^{s} q(i|s) = 1$ for all s. Then we have a binomial probability distribution

$$q(i|s) = \binom{s}{i} p^i (1-p)^{s-i}. \tag{3.37}$$

But now we have fuzzy probability $\bar{p}$ so we get the fuzzy binomial ([1],[2]). Hence

$$\bar{q}(i|s) = \binom{s}{i} \bar{p}^i (1-\bar{p})^{s-i}, \tag{3.38}$$

which is computed by α-cuts

$$\bar{q}(i|s)[\alpha] = \{ \binom{s}{i} p^i (1-p)^{s-i} | p \in \bar{p}[\alpha] \}, \tag{3.39}$$

for all α in $[0, 1]$. We will need this calculation in Chapter 8.

3.7 Figures

Some of the figures, graphs of fuzzy probabilities or certain fuzzy numbers, in the book are difficult to obtain so they were created using different methods. Many graphs were done first in Maple [10] and then exported to $LaTeX2_\epsilon$. We did these figures first in Maple because of the "implicitplot" command in Maple. Let us explain why this command was important in this book. Suppose $\overline{P}$ is a fuzzy probability we want to graph. Usually in this book we determine $\overline{P}$ by first calculating its α-cuts. Let $\overline{P}[\alpha] = [p_1(\alpha), p_2(\alpha)]$. So we get $x = p_1(\alpha)$ describing the left side of the triangular shaped fuzzy number $\overline{P}$ and $x = p_2(\alpha)$ describes the right side. On a graph we would have the x-axis horizontal and the y-axis vertical. α is on the y-axis between zero and one. Substituting y for α we need to graph $x = p_i(y)$, for $i = 1, 2$. But this is backwards, we usually have y a function of x. The "implicitplot" command allows us to do the correct graph with x a function of y when we have $x = p_i(y)$. The figures done in Maple and then exported to $LaTeX2_\epsilon$ are Figures 2.1-2.4, 3.1-3.3 and 12.1,12.2. Figures 2.5,2.6, 13.1-13.4,14.1, 15.1-15.2 and 16.1-16.2 were constructed in $LaTeX2_\epsilon$.

3.8 References

1. J.J.Buckley and E.Eslami: Uncertain Probabilities I: The Discrete Case, Soft Computing. To appear.

2. J.J.Buckley: Fuzzy Probabilities: New Approach and Applications, Physica-Verlag, Heidelberg, 2002. To appear.

3. Frontline Systems (www.frontsys.com).

4. R.V.Hogg and E.A.Tanis: Probability and Statistical Inference, Sixth Edition, Prentice Hall, Upper Saddle River, N.J., 2001.

5. G.J.Klir: Fuzzy Arithmetic with Requisite Constraints, Fuzzy Sets and Systems, 91(1997), pp. 147-161.

6. G.J.Klir and J.A.Cooper: On Constrainted Fuzzy Arithmetic, Proc. 5th Int. IEEE Conf. on Fuzzy Systems, Sept. 8-11, 1996, New Orleans, LA, pp. 1285-1290.

7. G.J.Klir and Y.Pan Y: Constrained Fuzzy Arithmetic: Basic Questions and Some Answers, Soft Computing, 2(1998), pp. 100-108.

8. V.Kreinovich, H.T.Nguyen, S.Ferson and L.Ginzburg: From Computation with Guaranteed Intervals to Computation with Confidence Intervals: A New Application of Fuzzy Techniques, Proc. 21st NAFIPS 2002, New Orleans, LA, June 27-29, 2002, pp. 418-422.

9. D.A.Menasce and V.A.F.Almeida: Capacity Planning for Web Performance, Prentice Hall, Upper Saddle River, N.J., 1998.

10. Maple 6, Waterloo Maple Inc., Waterloo, Canada.

11. Y.Pan and G.J.Klir: Bayesian Inference Based on Interval-Valued Prior Distributions and Likelihoods, J. of Intelligent and Fuzzy Systems, 5(1997), pp. 193-203.

12. Y.Pan and B.Yuan: Baysian Inference of Fuzzy Probabilities, Int. J. General Systems, 26(1997), pp. 73-90.

13. H.A.Taha: Operations Research, Fifth Edition, Macmillan, N.Y., 1992.

Chapter 4

Fuzzy Markov Chains

4.1 Introduction

The system we will look at has the properties : c parallel, independent and identical, servers; finite $(M, M > c)$ system capacity (in the queue and in the servers); and infinite calling source (where the customers come from). We will model the system as a fuzzy, finite, Markov chain. So in this chapter let us first briefly review the needed basic results from crisp (not fuzzy) finite Markov chains. Then we introduce: (1) fuzzy regular, finite, Markov chains; (2) fuzzy absorbing, finite, Markov chains; and (3) other fuzzy Markov chains needed in Chapter 14. Fuzzy regular Markov chains will be used throughout Chapters 5-10 and Chapters 13-17 but fuzzy absorbing, and other fuzzy Markov chains, will be needed only in Chapter 14. The next chapter deals with applying these results on fuzzy regular Markov chains to fuzzy queuing theory. Details on fuzzy Markov chains using fuzzy probabilities may be found in [1] and its application to fuzzy queuing theory is in [2]. Both topics are contained in the book [3].

A finite Markov chain has a finite number of possible states (outcomes) $S_1, S_2, \ldots, S_r$ at each step $n = 1, 2, 3 \ldots$, in the process. Let

$$p_{ij} = Prob(S_j \text{ at step } n+1 | S_i \text{ at step } n), \tag{4.1}$$

$1 \leq i, j \leq r$, $n = 1, 2, \ldots$. The p_{ij} are the transition probabilities which do not depend on n. The transition matrix $P = (p_{ij})$ is a $r \times r$ matrix of the transition probabilities. An important property of P is that the row sums are equal to one and each $p_{ij} \geq 0$. Let $p_{ij}^{(n)}$ be the probability of starting in state S_i and ending up in S_j after n steps. Define P^n to be the product of P n-times and it is well known that $P^n = (p_{ij}^{(n)})$ for all n. If $p^{(0)} = (p_1^{(0)}, \ldots, p_r^{(0)})$, where $p_i^{(0)} = $ the probability of initially being in state S_i, and $p^{(n)} = (p_1^{(n)}, \cdots, p_r^{(n)})$, where $p_i^{(n)} = $ the probability of being in state S_i after n steps, we know that $p^{(n)} = p^{(0)}P^n$.

4.2   Fuzzy Regular Markov Chains

We say that the Markov chain is regular if $P^k > 0$ for some k, which is $p_{ij}^{(k)} > 0$ for all i, j. This means that it is possible to go from any state S_i to any state S_j in k steps. A property of regular Markov chains is that powers of P converge, or $\lim_{n\to\infty} P^n = \Pi$, where the rows of Π are identical. Let w be the unique left eigenvalue of P corresponding to eigenvalue one, so that $w_i > 0$ all i and $\sum_{i=1}^r w_i = 1$. That is $wP = w$ for $1 \times r$ vector w. Each row in Π is equal to w and $p^{(n)} \to p^{(0)}\Pi = w$. After a long time, thinking that each step being a time interval, the probability of being in state S_i is w_i, $1 \le i \le r$, independent of the initial conditions $p^{(0)}$. In a regular Markov chain the process goes on forever jumping from state to state, to state, ...

Now proceed to a fuzzy finite, regular, Markov chain. All of the p_{ij} in the transition matrix P must be known. Suppose some of them are not known precisely and must be estimated and hence are uncertain. We substitute a fuzzy number $\overline{p}_{ij}$, as discussed in Chapter 3, for these uncertain p_{ij} producing a fuzzy transition $\overline{P}$. If some of the p_{ij} are known, like it is zero, we use these values but still write then as a fuzzy $\overline{p}_{ij}$.

In practice, from the data as discussed in Chapter 3, we first obtain fuzzy probabilities $\overline{p}(i)$ and $\overline{q}(l|s)$ from which we compute the fuzzy probabilities $\overline{p}_{ij}$. This computation is explained in the next chapter. So for this section we will assume that determining the $\overline{p}_{ij}$ from the $\overline{p}(i)$ and the $\overline{q}(l|s)$ has already been accomplished, and we will use triangular fuzzy numbers for $\overline{p}_{ij}$.

The uncertainty is in some of the p_{ij} values but not in the fact that the rows in the transition matrix must be discrete probability distributions (row sums equal one). So we now put the following restriction on the $\overline{p}_{ij}$ values: there are $p_{ij} \in \overline{p}_{ij}[1]$ so that $P = (p_{ij})$ is the transition matrix for a finite Markov chain (row sums one). At this point $\overline{P}$ is a $(M+1) \times (M+1)$ matrix with rows/columns numbered $0, 1, 2, ..., M$. The states are $S_0, ..., S_M$ where S_0 will mean no customers are in the system, S_1 is there is exactly one customer in the system, etc. We will need the following definitions for our restricted fuzzy matrix multiplication. Pick and fix an α in $[0, 1]$. Define $Dom[\alpha]$ as the set of all $p_{ij} \in \overline{p}_{ij}[\alpha]$, $0 \le i, j \le M$, so that if we form a transition matrix $P = (p_{ij})$ with these p_{ij} all the row sums equal one. Define $v = (p_{00}, p_{01}, ..., p_{MM})$. Row vector v is just all the p_{ij} in a transition matrix $P = (p_{ij})$. Then $Dom[\alpha]$ is all the vectors v, where the p_{ij} are in the alpha-cut of $\overline{p}_{ij}$ all i, j, so that P is the transition matrix for a finite Markov chain. In this section let us assume that P is regular.

For each $v \in Dom[\alpha]$ set $P = (p_{ij})$ and we get $P^n \to \Pi$. Let $\Gamma(\alpha) = \{w|wP = w, 0 < w_i < 1, w_0 + ... + w_M = 1, v \in Dom[\alpha]\}$. $\Gamma(\alpha)$ consists of all vectors w, which are the rows in Π, for all $v \in Dom[\alpha]$. Now the rows in $\overline{\Pi}$ will be all the same so let $\overline{w} = (\overline{w}_0 ...., \overline{w}_M)$ be a row in $\overline{\Pi}$. Also, let $\overline{w}_j[\alpha] = [w_{j1}(\alpha), w_{j2}(\alpha)]$, for $0 \le j \le M$. Then

$$w_{j1}(\alpha) = min\{w_j|w \in \Gamma(\alpha)\}, \tag{4.2}$$

and

$$w_{j2}(\alpha) = max\{w_j | w \in \Gamma(\alpha)\}, \qquad (4.3)$$

where w_j is the jth component in the vector w. The steady state fuzzy probabilities are : (1) $\overline{w}_0$ = the fuzzy probability of the system being empty; (2) $\overline{w}_1$ = the fuzzy probability of one customer in the system; etc.

In general, the solutions to equations (4.2) and (4.3) will be computationally difficult and one might consider using a genetic, or evolutionary, algorithm to get approximate solutions.

Let us go through an explanation of equations (4.2) and (4.3). Assume that the fuzzy transition matrix is 5×5 with rows numbered $0, 1, 2, 3, 4$ and the columns numbered $0, 1, 2, 3, 4$. Let $\overline{P} = (\overline{p}_{ij})$. Pick and fix some $\alpha \in [0, 1]$ and compute the intervals $I_{ij} = \overline{p}_{ij}[\alpha]$. Construct a 5×5 interval matrix $I = (I_{ij})$. Choose $a_{ij} \in I_{ij}$ so that $\sum_{j=0}^{4} a_{ij} = 1$ (row sums are one) for $i = 0, 1, 2, 3, 4$. Define $P = (a_{ij})$. Find w so that $wP = w$, $0 < w_i < 1$, $0 \le i \le 4$ and $w_0 + ... + w_4 = 1$. Define the process of randomly choosing the a_{ij} in I_{ij} so that all the row sums are equal to one, constructing P and then computing the left eigenvector w as process $\mathcal{RP}$. Then we can find the intervals $\overline{w}_i[\alpha] = [w_{i1}(\alpha), w_{i2}(\alpha)]$ as

$$w_{i1}(\alpha) = min\{w_i | \quad all \quad \mathcal{RP} \quad \}, \qquad (4.4)$$

and

$$w_{i2}(\alpha) = max\{w_i | \quad all \quad \mathcal{RP} \quad \}. \qquad (4.5)$$

Of course, we can not do "all $\mathcal{RP}$", so in practice we may do it 10,000 times to estimate the end points of these intervals. This is a purely random search, known to not be efficient, so we will adopt a genetic algorithm for this computation.

We have an alternate method of getting these fuzzy steady state probabilities. The 5×5 fuzzy transition matrix is still $\overline{P} = (\overline{p}_{ij})$, and $I_{ij} = \overline{p}_{ij}[\alpha]$ for some fixed alpha in $[0, 1]$. Set 5×5 interval matrix $I = (I_{ij})$. Choose $a_{ij} \in I_{ij}$, so that $\sum_{j=0}^{4} a_{ij} = 1$ (row sums are one), $i = 0, 1, ..., 4$. Define $A = (a_{ij})$ and $A^n \to \Pi$ with all rows in Π equal to $(w_0, w_1, ..., w_4)$. Define the process of randomly choosing the a_{ij} in I_{ij} so that all the row sums are one, constructing A and computing Π with the w_i, process $\mathcal{RP}$. Then we find the intervals $\overline{w}_i[\alpha] = [w_{i1}(\alpha), w_{i2}(\alpha)]$ as

$$w_{i1}(\alpha) = min\{w_i | \quad all \quad \mathcal{RP} \quad \}, \qquad (4.6)$$

and

$$w_{i2}(\alpha) = max\{w_i | \quad all \quad \mathcal{RP} \quad \}. \qquad (4.7)$$

The convergence of A^n to Π is easily done in Maple [5] for reasonably small matrices. Let $A1 = A^{10}$, $A2 = A1^{10}$,... We found that quite quickly the Ai converge to four decimal accuracy of the exact Π. Finding the rows $(w_0, ..., w_4)$ in the limit Π, to four place accuracy, for small M can be accomplished with Maple.

	$\alpha = 0$	$\alpha = 1$
$\overline{w}_0$	$[0.0067, 0.0243]$	0.0144
$\overline{w}_1$	$[0.0765, 0.1448]$	0.1117
$\overline{w}_2$	$[0.8339, 0.9152]$	0.8740

Table 4.1: Alpha-cuts of the Fuzzy Probabilities in Example 4.2.1

Example 4.2.1

We will work with a 3×3 fuzzy transition matrix $\overline{P}$ shown in equation (4.8). In equation (4.8) we have

$$\overline{P} = \begin{pmatrix} 0.3 & 0.2 & 0.5 \\ \overline{p}_{10} & \overline{p}_{11} & \overline{p}_{12} \\ 0 & \overline{p}_{21} & \overline{p}_{22} \end{pmatrix}, \tag{4.8}$$

$\overline{p}_{10} = (0.06/0.09/0.12)$, $\overline{p}_{11} = (0.26/0.27/0.28)$, $\overline{p}_{12} = (0.62/0.64/0.66)$, $\overline{p}_{21} = (0.06/0.09/0.12)$, and $\overline{p}_{22} = (0.88/0.91/0.94)$. Using the $\alpha = 0$ cut the interval matrix is

$$I = \begin{pmatrix} 0.3 & 0.2 & 0.5 \\ [0.06, 0.12] & [0.26, 0.28] & [0.62, 0.66] \\ 0 & [0.06, 0.12] & [0.88, 0.94] \end{pmatrix} \tag{4.9}$$

We have computed the $\alpha = 0$ and $\alpha = 1$ cuts of the fuzzy steady state probabilities and they are shown in Table 4.1. The $\alpha = 1$ cut is easy to use because we pick the vertex points of the triangular fuzzy numbers and we get a crisp transition matrix

$$P = \begin{pmatrix} 0.3 & 0.2 & 0.5 \\ 0.09 & 0.27 & 0.64 \\ 0 & 0.09 & 0.91 \end{pmatrix} \tag{4.10}$$

For this crisp P we easily find $P^n \to \Pi$ and the rows of Π give the $\alpha = 1$ cuts of the fuzzy steady state probabilities in Table 4.1.

For the $\alpha = 0$ cuts we did an exhaustive search since it is a small matrix with only five intervals. Given the accuracy of the data (two decimal places) in the interval matrix I, equation (4.9), we set up the search space Ω to be: (1) $p_{00} = 0.3$; (2) $p_{01} = 0.2$; (3) $p_{02} = 0.50$; (4) $p_{10} = 0.06, 0.07, ..., 0.12$; (5) $p_{11} = 0.26, 0.27, 0.28$; (6) $p_{12} = 0.62, 0.63, ..., 0.66$; (7) $p_{20} = 0$; (8) $p_{21} = 0.06, 0.07, ..., 0.12$; and (9) $p_{22} = 0.88, 0.89, ..., 0.94$. Any element in Ω where all the row sums are not one was rejected. These values of the p_{ij} make the crisp transition matrix P whose limit gives values for the w_i, whose min/max values produce the end points of the $\alpha = 0$ cut shown in Table 4.1. Naturally this exhaustive search can not be employed in larger matrices. For larger matrices we will employ a genetic algorithm in later chapters ([6],[7]). We did apply our genetic algorithm, and also the Premium Solver Platform V5.0 from Frontline Systems [4], to this problem and their answers agreed with those in Table 4.1. More details on our genetic algorithm and the Premium Solver are in Chapter 19.

4.3 Fuzzy Absorbing Markov Chains

First we will discuss the basic results for crisp absorbing Markov chains. We will call a state S_i absorbing if $p_{ii} = 1$, $p_{ij} = 0$ for $i \neq j$. Once in S_i you can never leave. Suppose there are k absorbing states, $1 \leq k < r$, and then we may rename the states (if needed) so that the transition matrix P can be written as

$$P = \begin{pmatrix} I & O \\ R & Q \end{pmatrix}, \tag{4.11}$$

where I is the $k \times k$ identity, O is the $k \times (r-k)$ zero matrix, R is $(r-k) \times k$ and Q is $(r-k) \times (r-k)$. The Markov chain is called an absorbing Markov chain if it has at least one absorbing state and from every non-absorbing state it is possible to reach some absorbing state in a finite number of steps. Assume the chain is absorbing and then we know that

$$P^n = \begin{pmatrix} I & O \\ SR & Q^n \end{pmatrix}, \tag{4.12}$$

where $S = I + Q + \cdots + Q^{n-1}$. Then $\lim_{n \to \infty} P^n = \Pi$ where

$$\Pi = \begin{pmatrix} I & O \\ R^* & O \end{pmatrix}, \tag{4.13}$$

for $R^* = (I - Q)^{-1} R$. Notice the zero columns in Π which implies that the probability that the process will eventually enter an absorbing state is one. The process eventually ends up in an absorbing state.

If $R = (r_{ij})$ and $Q = (q_{ij})$ we now assume that there is uncertainty in some of the r_{ij} and/or the q_{ij} values. We then substitute $\overline{r}_{ij}$ for r_{ij} and $\overline{q}_{ij}$ for q_{ij} and obtain $\overline{P}$ an absorbing fuzzy Markov chain. We now show that $\overline{P}^n \to \overline{\Pi}$ where

$$\overline{\Pi} = \begin{pmatrix} I & O \\ \overline{R}^* & O \end{pmatrix} \tag{4.14}$$

with $(r-k) \times k$ matrix $\overline{R}^* = (\overline{r}_{ij}^*)$. For any $v \in Dom[\alpha]$, P^n converges to the Π in equation (4.13) which implies that $\overline{Q}^n \to O$, the (crisp) zero matrix. Also, for any $v \in Dom[\alpha]$, $R^* = (I - Q)^{-1} R = (r_{ij}^*)$. Let $\overline{r}_{ij}^*[\alpha] = [r_{ij1}^*(\alpha), r_{ij2}^*(\alpha)]$. It follows that

$$r_{ij1}^*(\alpha) = min\{r_{ij}^* | v \in Dom[\alpha]\}, \tag{4.15}$$

and

$$r_{ij2}^*(\alpha) = max\{r_{ij}^*|v \in Dom[\alpha]\}. \tag{4.16}$$

To find the limit of $\overline{P}^n$, as $n \to \infty$, which is $\overline{\Pi}$, all we need to do is solve equations (4.15) and (4.16) for the α-cuts of the $\overline{r}_{ij}^*$ in $\overline{R}^*$ in equation (4.14).

Example 4.3.1

Let

$$P = \begin{pmatrix} 1 & 0 & 0 \\ 0.2 & 0.6 & 0.2 \\ 0.4 & 0.3 & 0.3 \end{pmatrix} \tag{4.17}$$

have one absorbing state. Substitute $\overline{p}_{ij}$ for p_{ij}, not in the first row, expressing the uncertainty in the p_{ij} values. Then $\overline{P}^n \to \overline{\Pi}$ where $\overline{\Pi}$ is the crisp matrix

$$\overline{\Pi} = \begin{pmatrix} 1 & 0 & 0 \\ 1 & 0 & 0 \\ 1 & 0 & 0 \end{pmatrix}, \tag{4.18}$$

because $\overline{Q}^n \to O$ and the $\overline{r}_{i1}^*$ must equal crisp one because the row sums are one.

Let us now discuss how Example 4.3.1 will be used in Chapter 14. First assume that P is 5×5 with the rows, and columns, labeled "$0,1,2,3,4$". The absorbing state is state "4" so we rearrange the row and columns so that they are labeled "$4,0,1,2,3$". Let $P = (p_{ij})$. Then $p_{00} = 1$ and $p_{0j} = 0$ for $j = 1,2,3,4$. All row sums are equal to one. It follows that $P^n \to \Pi$ where all the rows in Π are equal to $(1,0,0,0,0)$. This limit gives the steady state probabilities. This limit is the same no matter what the p_{ij} values as long as the row sums must be one and the first row in P is $(1,0,0,0,0)$. Since we moved state "4" up to first place, with respect to the original labeling, the fuzzy steady state probability is $(0,0,0,0,1)$.

Our second application is to 11×11 P whose rows/columns are labeled "0,1,2,...,10". The absorbing state is "10". Rearrange the rows/columns so that they are labeled "10,0,1,2,...,9". All row sums are one. Now $p_{00} = 1$ and $p_{0j} = 0$ for $j = 1,2,3,...,10$. Then $P^n \to \Pi$ with all the rows in Π equal to $(1,0,0,0,0,0,0,0,0,0,0)$. This limit is the steady state probabilities and will be the same for all p_{ij} values if the row sums are one and the first row in P is $(1,0,0,0,0,0,0,0,0,0,0)$. With respect to the original labeling of the states the fuzzy steady state probability is $(0,0,0,0,0,0,0,0,0,0,1)$.

	$\alpha = 1$	$\alpha = 0$
$\overline{r}_{11}^*$	0.3404	[0.2021.0.4787]
$\overline{r}_{12}^*$	0.6596	[0.5213,0.7979]
$\overline{r}_{21}^*$	0.4681	[0.3404,0.5979]
$\overline{r}_{22}^*$	0.5319	[0.4042,0.6569]

Table 4.2: Alpha-cuts of the Fuzzy Numbers $\overline{r}_{ij}^*$ in Example 4.3.2

Example 4.3.2

We next consider two absorbing states with transition matrix

$$P = \begin{pmatrix} 1 & 0 & 0 & 0 \\ 0 & 1 & 0 & 0 \\ 0.2 & 0.5 & 0 & 0.3 \\ 0.4 & 0.4 & 0.2 & 0 \end{pmatrix}. \tag{4.19}$$

Substitute $\overline{r}_{ij}$ for r_{ij} and $\overline{q}_{ij}$ for q_{ij} where $\overline{r}_{11} = (0.1/0.2/0.3)$, $\overline{r}_{12} = (0.4/0.5/0.6)$, $\overline{r}_{21} = (0.3/0.4/0.5)$, $\overline{r}_{22} = (0.3/0.4/0.5)$, $\overline{q}_{12} = (0.2/0.3/0.4)$ and $\overline{q}_{21} = (0.1/0.2/0.3)$. First we determine the r_{ij}^* values in terms of the r_{ij} and the q_{ij}. The result is $r_{11}^* = (r_{11} + r_{21}q_{12})/T$, $r_{12}^* = (r_{12} + r_{22}q_{12})/T$, $r_{21}^* = (r_{21} + r_{11}q_{21})/T$ and $r_{22}^* = (r_{22} + r_{12}q_{21})/T$ where $T = 1 - q_{21}q_{12}$. We may solve equations (4.15) and (4.16) by solving a non-linear optimization problem using the "Premium Solver Platform V5.0" from Frontline Systems [4]. For example, for the alpha equal zero cut $r_{21_2}^*$, the right end point of the interval for r_{21}^*, would be $max(r_{21} + r_{11}q_{21})/T$ subject to: $0.1 \leq r_{11} \leq 0.3$, $0.4 \leq r_{12} \leq 0.6$, $0.3 \leq r_{21} \leq 0.5$, $0.3 \leq r_{22} \leq 0.5$, $0.2 \leq q_{12} \leq 0.4$, $0.1 \leq q_{21} \leq 0.3$, $r_{11} + r_{12} + q_{12} = 1$ and $r_{21} + r_{22} + q_{21} = 1$. The results for the $\alpha = 0$ cuts are shown in Table 4.2.

4.4 Other Fuzzy Markov Chains

In Chapter 14 we will have a finite fuzzy Markov chain which is not regular nor is it absorbing. If $\overline{P}$ is the fuzzy transition matrix for such a fuzzy Markov chain we still obtain $\overline{P}^n \rightarrow \overline{\Pi}$ as $n \rightarrow \infty$. Let us look at an example of this type of $\overline{P}$.

Example 4.4.1

Consider

$$P = \begin{pmatrix} 0 & 0.4 & 0.3 & 0.2 & 0.1 \\ 0 & 0.16 & 0.36 & 0.26 & 0.22 \\ 0 & 0.064 & 0.24 & 0.32 & 0.376 \\ 0 & 0 & 0.064 & 0.24 & 0.696 \\ 0 & 0 & 0 & 0.064 & 0.936 \end{pmatrix}. \tag{4.20}$$

This is, for example, the $\alpha = 1$ cut of the fuzzy transition matrix. In this case the states of the system are numbered "0,1,2,3,4". So the rows of P are labeled "0,1,2,3,4" and the columns of P are also labeled "0,1,2,3,4". We see that it is impossible to go the state labeled zero. Notice that the row sums in this P all equal one. Clearly, P does not represent an absorbing Markov chain. Also, P^n will always have its first column all zeros, so P is not the transition matrix for a regular Markov chain.

Partition P as follows

$$P = \begin{pmatrix} 0 & R \\ 0 & A \end{pmatrix}, \tag{4.21}$$

where R is 1×4 and A is 4×4. Now we see that

$$P^n = \begin{pmatrix} 0 & RA^{n-1} \\ 0 & A^n \end{pmatrix}. \tag{4.22}$$

A is a 4×4 transition matrix for a regular Markov chain so $A^n \to \Pi$ as $n \to \infty$. Let the rows in Π be all equal to $(w_1, w_2, w_3, w_4) = w$. Also $RA^{n-1} \to R\Pi = w$. So P^n converges to a 5×5 matrix whose rows are all equal to $(0, w_1, w_2, w_3, w_4)$ and the w_i are the steady state probabilities. For the numbers given in the first P we calculate the steady state probabilities as $[0, 0.0005, 0.0070, 0.0800, 0.9125]$ using Maple [5].

The application of Example 4.4.1 will be in Chapter 14 for a 5×5 , and a 11×11, transition matrix P. In both cases the first column in $\overline{P}$ will be all zeros. If the rows/columns are labeled "$0, 1, ..., n-1$", then it is impossible to enter state "0". Suppose P is $n \times n$, row sums all equal to one and the first column all zeros. We get $P = (p_{ij})$ from $p_{ij} \in \overline{p}_{ij}[\alpha]$, $0 \le i \le n-1$. $1 \le j \le n-1$ all α. Let P_0 be the $(n-1) \times (n-1)$ submatrix of P obtained by deleting the first row and column of P. Then P_0 will be a transition matrix for a regular Markov chain and $P_0^n \to \Pi$ with all the rows in Π equal to $(w_1, ..., w_{n-1})$. It follows that the steady state probabilities for P will be $(0, w_1, ..., w_{n-1})$. Now let $(\overline{w}_1, ..., \overline{w}_{n-1})$ be the fuzzy steady state probabilities for $\overline{P}_0$ and then the fuzzy steady state probabilities for $\overline{P}$ are $(0, \overline{w}_1, ..., \overline{w}_{n-1})$.

4.5   References

1. J.J.Buckley and E.Eslami: Fuzzy Markov Chains: Uncertain Probabilities, MathWare and Soft Computing, 9(2002), pp. 33-41.

2. J.J.Buckley and E.Eslami: Uncertain Probabilities I: The Discrete Case, Soft Computing. To appear.

3. J.J.Buckley: Fuzzy Probabilities: New Approach and Applications, Physica-Verlag, Heidelberg, 2003.

4. Frontline Systems (www.frontsys.com).

5. Maple 6, Waterloo Maple Inc., Waterloo, Canada.

6. X.Zheng, K.Reilly and J.J.Buckley: Applying Genetic Algorithms to Fuzzy Probability-Based Web Planning Models, Proceedings ACMSE, Savannah, Ga, 2003, pp. 241-245.

7. X.Zheng, K.Reilly and J.J.Buckley: Comparing Genetic Algorithms and Exhaustive Methods Used in Optimization Problems for Fuzzy Probability Based Web Planning Models, Proceedings 2003 Int. Conf. on AI, June 23-26, 2003, Las Vegas, Nevada. To appear.

Chapter 5

Fuzzy Queuing Theory

5.1 Introduction

In this chapter we apply the results in Chapter 4 on fuzzy finite, regular, Markov chains to fuzzy queuing theory. The crisp system is as explained in Chapter 4: c parallel, independent, and identical servers, $M \geq c$ is system capacity ($M - c$ maximum queue size), and a calling source so large that we may approximate it as an infinite calling source. We will model the crisp queuing system as a finite, regular, Markov chain. The other finite Markov chains discussed in Chapter 4 will only appear in Chapters 14.

5.2 Queuing Theory

We first need to explain how the system can change at the end of each time interval δ. System changes can occur only at the end of a time interval δ. This time interval may be one second, one minute, one hour, etc. During a time interval δ: (1) customers may arrive at the system but are only allowed into the system at the end of the time interval; (2) customers may leave the servers but are allowed to return to the calling source only at the end of the time interval; (3) at the end of the time interval all customers in queue (in the system but not in the servers) are allowed to fill the empty servers; and (4) all customers who arrived are allowed into the system to fill empty servers or go into queue up to capacity M with all others turned away to return to the calling source. System changes can occur only at times $t = \delta$, 2δ, 3δ, ...

Let $p(i)$ be the probability that i customers arrive at the system during a time interval δ, $i = 0, 1, 2, 3, ...$ Then $\sum_{i=0}^{\infty} p(i) = 1$. Next let $q(l|s)$ be the probability that, during a time interval δ, l customers in the servers complete service and are waiting to return to the calling source at the end of the time interval, given that s servers are full of customers at the start of the time period, for $l = 0, 1, 2, ..., s$ and $s = 0, 1, 2, 3, ...c$. Then $\sum_{l=0}^{s} q(l|s) = 1$ for each s. Next we construct the transition matrix P. The rows of P are labeled $0, 1, 2, 3, ..., M$ representing the state of the system at the start of the time period and the columns of P are labeled $0, 1, 2, 3, ..., M$ representing the state of the system at the beginning of the next period. To see how we compute the p_{ij} in P let us look at the following example.

Example 5.2.1

Let $c = 2$ and $M = 4$. Then $P = (p_{ij})$, a 5×5 matrix with probabilities p_{ij}, $0 \le i, j \le 4$. Let us compute the p_{2j} in the third row: (1) $p_{20} = p(0)q(2|2)$; (2) $p_{21} = p(0)q(1|2)+p(1)q(2|2)$; (3) $p_{22} = p(0)q(0|2)+p(1)q(1|2)+p(2)q(2|2)$; (4) $p_{23} = p(1)q(0|2) + p(2)q(1|2) + p(3)q(2|2)$; and (5) $p_{24} = p^*(2)q(0|2) + p^*(3)q(1|2)+p^*(4)q(2|2)$ where $p^*(k) = \sum_{i=k}^{\infty} p(i)$. Notice that $\sum_{j=0}^{4} p_{2j} = 1$ so P is the transition matrix for a finite Markov chain. Also, $P^2 > 0$ so it is regular. The whole transition matrix is presented in Table 8.1 in Chapter 8. Table 6.2 in Chapter 6 is the transition matrix when $c = 1$ and $M = 4$. Also, the transition matrix for $c = 4$ and $M = 10$ is discussed in Chapter 10.

We see from Example 5.2.1 that P will be the transition matrix for a regular, finite, Markov chain. Then $P^n \to \Pi$ where each row in Π is $w = (w_0, ..., w_M)$ and $w_i > 0$, $w_0 + ... + w_M = 1$, $wP = w$. Also $p^{(n)} \to w$ so, after a long time, the probability of being in state S_j is w_j, $0 \le j \le M$, independent of the initial conditions.

5.3   Fuzzy Queuing Theory

The $p(i)$ and the $q(l|s)$ have to be known exactly. Many times they have to be estimated and if the queuing system is in the planning stage, these probabilities will be estimated by experts. So we assume there is uncertainty in some of the $p(i)$ and $q(l|s)$ values which implies uncertainty in the p_{ij} values in P. If a $p_{ij} = 0$ in P, then we assume there is no uncertainty in this value. We model the uncertainty by substituting fuzzy numbers $\overline{p}_{ij}$ for all the non-zero p_{ij} in P. Of course. if some p_{ij} value is known, say $p_{24} = 0.15$ with no uncertainty, then $\overline{p}_{24} = 0.15$ also. We then obtain a fuzzy transition matrix $\overline{P} = (\overline{p}_{ij})$ so that if $p_{ij} = 0$ then $\overline{p}_{ij} = 0$ and if $0 < p_{ij} < 1$, then $0 < \overline{p}_{ij} < 1$ also. We know from Chapter 4, under restricted fuzzy , that $\overline{P}^n \to \overline{\Pi}$ where each row in $\overline{\Pi}$ is $\overline{w} = (\overline{w}_0, ..., \overline{w}_M)$ and the $\overline{w}_i$ give the fuzzy steady state probabilities of the fuzzy queuing system.

We presented three solution methods for the fuzzy steady state probabilities in the previous chapter. More details on the computational methods/problems starts in the next chapter and is in Chapter 19.

Chapter 6

Computations: One Sever

6.1 Introduction

Starting with the fuzzy probabilities $\overline{p}(i)$ and $\overline{p}$, we wish to propagate these uncertainties through the model to finally compute fuzzy numbers for server utilization $(\overline{U})$, average server throughput $(\overline{X})$, average number of requests in the server $(\overline{N})$ and average response time $(\overline{R})$ ([4], Chapter 8). These fuzzy numbers will display the uncertainties in these system descriptors. The base of these fuzzy numbers, the $\alpha = 0$ cut, is like a 99% confidence interval.

We assume c parallel, independent and identical servers, M system capacity and infinite calling source (potential customer base). We will use $M = 4$ in this chapter. The calculation proceeds in three steps: (1) calculate the fuzzy probabilities $\overline{p}_{ij}$ in the $(M + 1) \times (M + 1)$ transition matrix $\overline{P}$; (2) find the steady state fuzzy probabilities $\overline{w}_i$, $0 \le i \le M$; and (3) determine the fuzzy numbers $\overline{U}$, $\overline{X}$, $\overline{N}$ and $\overline{R}$. The calculations are based on [1] and Chapter 3 in [2]. To keep the discussion fairly simple let $M = 4$ and we start with $c = 1$, then consider $c = 2$ later in Chapter 8. The data we will use in the following examples, plus to be used in the example in the next section, is given in Table 6.1 for the fuzzy arrival probabilities. Notice that we have assumed that $\overline{p}(i) = 0$ for $i > 7$. The other fuzzy probability that we need is $\overline{p} = (0.3/0.4/0.5)$, which is the fuzzy probability that a customer leaves the server during time interval δ, given that the customer was in the server at the start of the time interval. All of these fuzzy numbers have been taken as triangular fuzzy numbers.

6.2 Calculations

We now go through sample computations in the three steps outlined above when there is one server.

$\overline{p}$	Fuzzy Probability
$\overline{p}(0)$	$(0.07/0.1/0.13)$
$\overline{p}(1)$	$(0.26/0.3/0.34)$
$\overline{p}(2)$	$(0.17/0.2/0.23)$
$\overline{p}(3)$	$(0.07/0.1/0.13)$
$\overline{p}(4)$	$(0.07/0.1/0.13)$
$\overline{p}(5)$	$(0.07/0.1/0.13)$
$\overline{p}(6)$	$(0.03/0.05/0.07)$
$\overline{p}(7)$	$(0.03/0.05/0.07)$

Table 6.1: Fuzzy Probabilities for Arrivals

Previous State	Future State				
	0	1	2	3	4
0	$p(0)$	$p(1)$	$p(2)$	$p(3)$	$p^*(4)$
1	$p(0)p$	$p(1)p+$	$p(2)p+$	$p(3)p+$	$p^*(4)p+$
		$p(0)(1-p)$	$p(1)(1-p)$	$p(2)(1-p)$	$p^*(3)(1-p)$
2	0	$p(0)p$	$p(1)p+$	$p(2)p+$	$p^*(3)p+$
			$p(0)(1-p)$	$p(1)(1-p)$	$p^*(2)(1-p)$
3	0	0	$p(0)p$	$p(1)p+$	$p^*(2)p+$
				$p(0)(1-p)$	$p^*(1)(1-p)$
4	0	0	0	$p(0)p$	$p^*(1)p+(1-p)$

Table 6.2: The Transition Matrix P for $c = 1$ and $M = 4$

Step 1

We determine the fuzzy $\overline{p}_{ij}$ in the 5×5 fuzzy transition matrix $\overline{P}$. The crisp transition matrix is shown in Table 6.2. In Table 6.2 $p^*(i)$ stands for $\sum_{k=i}^{7} p(k)$, $i = 1, 2, 3, 4$. Two sample calculations would be for $\overline{p}_{13}$ and $\overline{p}_{34}$.

From Table 6.2 we see that $\overline{p}_{13} = \overline{p}(3)\overline{p} + \overline{p}(2)(1 - \overline{p})$. However, we do not substitute fuzzy numbers in for $\overline{p}(3)$, $\overline{p}(2)$ and $\overline{p}$ into the above equation, then do the multiplication and addition, to get $\overline{p}_{13}$. We determine $\overline{p}_{13}$ from its α-cuts ([1],[2])

$$\overline{p}_{13}[\alpha] = \{p(3)p + p(2)(1 - p)| \quad \mathbf{S}_1 \quad \}, \qquad (6.1)$$

for all $\alpha \in [0, 1]$, where $\mathbf{S}_1$ is the statement " $p(i) \in \overline{p}(i)[\alpha]$, $0 \le i \le 7$, $\sum_{i=0}^{7} p(i) = 1$, $p \in \overline{p}[\alpha]$". We first give the result for any α but do the numerical calculation only for $\alpha = 0$. We calculate fuzzy probabilities using "restricted" fuzzy arithmetic (Sections 3.5 and 3.6 of Chapter 3). There is uncertainty in the values of the $p(i)$, so we pick them in the α-cuts of the fuzzy probabilities, but there is no uncertainty that we have a discrete probability distribution over the set $\{0, 1, 2, ..., 7\}$, which is the number of customers arriving per unit time, so we constrain the sum to be one.

Now we determine $\bar{p}_{13}[0]$. Let $f = p(3)p + p(2)(1 - p)$. We see that the partial of f with respect to p is $p(3) - p(2)$ which is negative for all $p(3) \in \bar{p}(3)[0] = [0.07, 0.13]$ and for all $p(2) \in \bar{p}(2)[0] = [0.17, 0.23]$. Clearly f is an increasing function of both $p(2)$ and $p(3)$. So the left end point of $\bar{p}_{13}[0]$ should be $(0.17)(0.5) + (0.07)(0.5) = 0.12$ using the smallest values for $p(2)$ and $p(3)$ with the largest value for p. Similarly, the right end point should be $(0.23)(0.7) + (0.13)(0.3) = 0.20$. However, we did not check to see if these values for $p(2)$ and $p(3)$ were "feasible".

Values $v_1 = p(2)$ in $\bar{p}(2)[0]$ and $v_2 = p(3)$ in $\bar{p}(3)[0]$ are feasible if there exist $p(i) \in \bar{p}(i)[0]$, $i = 0, 1, 4, 5, 6, 7$, so that $p(0) + p(1) + v_1 + v_2 + p(4) + ... + p(7) = 1$. We now check and see that $v_1 = 0.17, v_2 = 0.07$ are feasible and $v_1 = 0.23, v_2 = 0.13$ are also feasible. We conclude that

$$\bar{p}_{13}[0] = [0.12, 0.20]. \tag{6.2}$$

In a similar manner we find $\bar{p}_{ij}[0]$ for $ij = 00, 01, 02, 03, 04, 10, 11, 12, 13, 21,$ $22, 23, 32, 33, 43$.

Next we turn to finding $\bar{p}_{34}[0]$. The method used also applies to $\bar{p}_{ij}[0]$ for $ij = 14, 24, 44$. From Table 6.2 we see

$$\bar{p}_{34} = [\sum_{i=2}^{7} \bar{p}(i)]\bar{p} + [\sum_{i=1}^{7} \bar{p}(i)](1 - \bar{p}), \tag{6.3}$$

which will be evaluated using α-cuts. The alpha equal zero cut is

$$\bar{p}_{34}[0] = \{[\sum_{i=2}^{7} p(i)]p + [\sum_{i=1}^{7} p(i)](1 - p)| \quad \mathbf{S_1} \quad \}, \tag{6.4}$$

where we put $\alpha = 0$ in "$\mathbf{S_1}$". Let f be the expression to be evaluated in the right side of equation (6.4). Then we may simplify f to $\sum_{i=2}^{7} p(i) + p(1)(1-p)$. From this we see that f is a decreasing function of p. Clearly f is an increasing function of $p(i)$, $1 \leq i \leq 7$. So, to find the left (right) end point we would like to use the left (right) end point of $\bar{p}(i)[0]$ for $p(i)$, $1 \leq i \leq 7$, and $p = 0.5$ ($p = 0.3$). But these values of $p(i)$, $1 \leq i \leq 7$, are not feasible. If we use the left (right) end points of $\bar{p}(i)[0]$ for $p(i)$, $1 \leq i \leq 7$, there is no value of $p(0) \in \bar{p}(0)[0]$ so that the sum of the $p(i)$ equals one. Therefore, we need to compute $\bar{p}_{34}[0]$ differently.

It turns out that finding the end points of the interval $\bar{p}_{34}[0]$ is accomplished by solving a linear programming problem. Since f is a decreasing function of p we can use $p = 0.5$ ($p = 0.3$) for the left (right) end point. Then the objective functions are, for the left end point, a minimization problem

$$OBJ_1 = (0.5)p(1) + \sum_{i=2}^{7} p(i), \tag{6.5}$$

and for the right end point, a maximization problem

$$OBJ_2 = (0.7)p(1) + \sum_{i=2}^{7} p(i). \tag{6.6}$$

Let $\overline{p}(i)[0] = [p_{i1}, p_{i2}]$, $0 \le i \le 7$. Subject to constraints

$$p_{i1} \le p(i) \le p_{i2}, 0 \le i \le 7, \tag{6.7}$$

and

$$p(0) + p(1) + ... + p(7) = 1. \tag{6.8}$$

Using Maple [5] to solve these two linear programming problems we obtained $OBJ_1 = 0.700$ and $OBJ_2 = 0.852$. Hence $\overline{p}_{34}[0] = [0.700, 0.852]$.

Now assume that we have calculated $\overline{p}_{ij}[\alpha]$, $0 \le i, j \le 4$, for selected values of alpha in $[0, 1]$.

Step 2

Here we determine the fuzzy steady state probabilities $\overline{w}_i$, $0 \le i \le 4$. Three methods of finding alpha-cuts of these fuzzy probabilities were presented in Chapter 4. We will use a genetic algorithm to estimate the end points of the alpha-cuts of the fuzzy steady state probabilities based on the left eigenvector method given in Chapter 4, which was the first method presented in that chapter. Details of our genetic algorithm and how it was adapted to our different optimization problems is in Chapter 19. We also will use the Premium Solver Platform V5.0 from Frontline Systems [3] to estimate the end points of the $\alpha-$cuts of the fuzzy steady state probabilities for these 5×5 matrices. More details of these two solution methods applied to finding the fuzzy steady state probabilities is in Chapter 19.

We now assume that we have reasonable estimates of certain alpha-cuts of the fuzzy steady state probabilities.

Step 3

The last thing to do in this section is get the fuzzy numbers $\overline{U}$, $\overline{X}$, etc. All we need are the $\overline{w}_i$ and $\overline{p}$. To motivate the fuzzy calculation we will first present the crisp definition. The crisp probabilities are w_i and p.

The crisp definition of U is

$$U = \sum_{i=1}^{M} w_i, \tag{6.9}$$

which equals $1 - w_0$. $U \times 100$ gives the percentage of time we expect the server to be busy. In the fuzzy case

$$\overline{U} = \sum_{i=1}^{M} \overline{w}_i, \tag{6.10}$$

which may, or may not, be $1 - \overline{w}_0$, and is evaluated by α-cuts

$$\overline{U}[\alpha] = \{\sum_{i=1}^{M} w_i | \quad \mathbf{S}_2 \quad \}, \tag{6.11}$$

for all α, where $\mathbf{S}_2$ is "$w_i \in \overline{w}_i[\alpha]$, $0 \leq i \leq M$, $w_0 + ...w_M = 1$". We can find the end points of the interval $\overline{U}[\alpha]$ by solving a linear programming problem, using the usual solution method, which will be discussed further in Step 3 of Chapter 7. We will need to change this when $c > 1$.

N is just the expected number of customers in the system

$$N = \sum_{k=0}^{M} k w_k, \tag{6.12}$$

and $\overline{N}$ is determined by its α-cuts

$$\overline{N}[\alpha] = \{\sum_{k=0}^{M} k w_k | \quad \mathbf{S}_2 \quad \}. \tag{6.13}$$

The end points of the interval $\overline{N}[\alpha]$ may also be found by solving a linear programming problem, also discussed in more detail in Step 3 of Chapter 7.

X is the expected number of customers leaving the system per time period δ. The crisp X is $(1)(p)(U) + (0)$(probability no one leaves the system). Fuzzy X is $\overline{X} = \overline{p}\,\overline{U}$ and its α-cuts are

$$\overline{X}[\alpha] = \{(p)(U) | p \in \overline{p}[\alpha], \mathbf{S}_2\}. \tag{6.14}$$

$\overline{U}$ is from equation (6.11) with $M = 4$. End points of the interval $\overline{X}[\alpha]$ may also be obtained by solving a linear programming problem as discussed in Step 3 in Chapter 7. This calculation becomes more complicated if $c > 1$.

R is the expected response of the system. For example, if the average number customers in the system is 20 (assume $M \geq 20$), and the expected number of customers leaving the system per time period (throughput) is 4 (assume $c \geq 4$), then $R = \frac{20}{4} = 5$ time periods δ. So, in the fuzzy case

$$\overline{R} = \frac{\overline{N}}{\overline{X}}. \tag{6.15}$$

6.3 References

1. J.J.Buckley and E.Eslami: Uncertain Probabilities I: The Discrete Case, Soft Computing. To appear.

2. J.J.Buckley: Fuzzy Probabilities: New Approach and Applications, Physica-Verlag, Heidelberg, 2003.

3. Frontline Systems (www.frontsys.com).

4. D.A.Menasce and V.A.F.Almeida: Capacity Planning for Web Performance, Prentice Hall, Upper Saddle River, N.J., 1998.

5. Maple 6, Waterloo Maple Inc., Waterloo, Canada.

Chapter 7

Example: One Sever

7.1 Introduction

In this chapter we will do the computations discussed in Chapter 6. This is a small system comparable to examples in [2].

So $c = 1$, $M = 4$, infinite calling source, the fuzzy probabilities of arrival are in Table 6.1 of Chapter 6, and $\overline{p} = (0.3/0.4/0.5)$ is the fuzzy probability that a customer leaves a busy server during the time interval δ. The crisp transition matrix is in Table 6.2 of Chapter 6. We will only compute the $\alpha = 0$ and $\alpha = 1$ cut of the fuzzy steady state probabilities and the fuzzy numbers ($\overline{U}$, $\overline{R}$, etc.) of system performance.

7.2 Computations

Step 1

The first thing to do is to find the alpha equal zero cut of the fuzzy transition probabilities $\overline{p}_{ij}$ in the fuzzy transition matrix $\overline{P}$. These computations were discussed in Section 6.2 of Chapter 6. The intervals were put into interval matrix $I = (I_{ij})$. The result of these calculations are shown in Table 7.1.

Step 2

Let us now discuss how we estimated the end points of the $\alpha = 0$ cuts of the fuzzy steady state probabilities. We did an exhaustive search since the interval matrix is a relatively small matrix with only 19 intervals. Given the accuracy of the data (number of decimal places) in the interval matrix I, Table

I	Alpha Zero Cut	Alpha One Cut
I_{00}	$[0.07, 0.13]$	0.1
I_{01}	$[0.26, 0.34]$	0.3
I_{02}	$[0.17, 0.23]$	0.2
I_{03}	$[0.07, 0.13]$	0.1
I_{04}	$[0.2, 0.4]$	0.3
I_{10}	$[0.021, 0.065]$	0.04
I_{11}	$[0.127, 0.235]$	0.18
I_{12}	$[0.215, 0.307]$	0.26
I_{13}	$[0.12, 0.20]$	0.16
I_{14}	$[0.25, 0.47]$	0.36
I_{20}	0	0
I_{21}	$[0.021, 0.065]$	0.04
I_{22}	$[0.127, 0.235]$	0.18
I_{23}	$[0.215, 0.307]$	0.26
I_{24}	$[0.415, 0.619]$	0.52
I_{30}	0	0
I_{31}	0	0
I_{32}	$[0.021, 0.065]$	0.04
I_{33}	$[0.127, 0.235]$	0.18
I_{34}	$[0.700, 0.852]$	0.78
I_{40}	0	0
I_{41}	0	0
I_{42}	0	0
I_{43}	$[0.021, 0.065]$	0.04
I_{44}	$[0.935, 0.979]$	0.96

Table 7.1: Alpha Cuts of Fuzzy Transition Matrix, One Server

$\overline{w}$	Alpha Zero Cut	Alpha One Cut
$\overline{w}_0$	$[0.0000^*, 0.0000^*]$	0.0000^*
$\overline{w}_1$	$[0.0000^*, 0.0006]$	0.0001
$\overline{w}_2$	$[0.0006, 0.0070]$	0.0023
$\overline{w}_3$	$[0.0236, 0.0804]$	0.0471
$\overline{w}_4$	$[0.9122, 0.9758]$	0.9504

Table 7.2: Alpha Cuts of Fuzzy Steady State Probabilities, One Server

7.1, we set up the search space Ω to be: (1) $p_{00} = 0.07, 0.08, ..., 0.13$; (2) $p_{01} = 0.26, 0.27, ..., 0.34$; (3) $p_{02} = 0.17, 0.18, ..., 0.23$; (4) $p_{03} = 0.07, 0.08, ..., 0.13$; (5)$p_{04} = 0.2, 0.3, 0.4$; (6) $p_{10} = 0.021, 0.022, ..., 0.065$; (7) $p_{11} = 0.127, 0.128,$ $..., 0.235$; (8) $p_{12} = 0.215, 0.216, ..., 0.307$; (9) $p_{13} = 0.12, 0.13, ..., 0.20$; (10) $p_{14} = 0.25, 0.26, ..., 0.47$, etc. Any element in Ω where all the row sums are not one was rejected. These values of the p_{ij} make the crisp transition matrix P whose limit gives values for the w_i, whose min/max values produce the end points of the $\alpha = 0$ cut shown in Table 7.2. We took P to a sufficiently high power to guarantee 5 decimal accuracy, and then rounded our results off to four decimal places for Table 7.2. The computer time for this calculation was not too great; letting it run over night we would have the answers in the morning. Naturally this exhaustive search can not be employed with large matrices ([3],[4]).

We also estimated the end points of $\overline{w}_i[0]$, $0 \leq i \leq 4$, using a genetic algorithm. We had ten similar genetic algorithms, one for each of the two end points of the five intervals. Each algorithm ran for 100,000 generations. The results were the same as those in Table 7.2 to four decimal places. In Table 7.2 0.0000^* means it was zero rounded to four decimal places. Finally, we also employed the Premium Solver Platform from Frontline Systems [1] whose results agreed with those in Table 7.2

Step 3

Finally, using $\overline{p}$ and the $\overline{w}_i$ we may find the $\alpha = 0$ and $\alpha = 1$ cuts of $\overline{U}, \overline{N}, \overline{X}$ and $\overline{R}$ whose fuzzy expressions were presented in Section 6.2 of Chapter 6. The results are given in Table 7.3. In Table 7.3 1.0000^* means it was one when rounded off to four decimal places.

Let us now go through some of the details on how we got these α-cuts in Table 7.3. We compute the alpha-cuts of $\overline{U}$ from equation (6.11) of Chapter 6 by solving a linear programming problem. Let $\overline{w}_i[0] = [w_{i1}(0), w_{i2}(0)]$, $0 \leq i \leq 4$, which are obtained from Table 7.2. Let $\overline{U}[0] = [u_1(0), u_2(0)]$. The objective function is

$$max/min[w_1 + ... + w_4], \qquad (7.1)$$

	Alpha Zero Cut	Alpha One Cut
$\overline{U}$	$[1.0000^*, 1.0000^*]$	1.0000^*
$\overline{N}$	$[3.9038, 3.9752]$	3.94786
$\overline{X}$	$[0.3000, 0.5000]$	0.4000
$\overline{R}$	$[7.8076, 13.2513]$	9.8697

Table 7.3: Alpha Cuts of $\overline{U}, \overline{N}, \overline{X}, \overline{R}$, One Server

subject to constraints

$$w_{i1}(0) \le w_i \le w_{i2}(0), i = 0, ..., 4, w_0 + ... + w_4 = 1. \qquad (7.2)$$

The result is in Table 7.3. The solution to the min (max) problem gives $u_1(0)$ $(u_2(0))$.

Alpha-cuts of $\overline{N}$ are given in equation (6.13) of Chapter 6. Let $\overline{N}[0] = [n_1(0), n_2(0)]$. Now we need to solve the following linear programming problem

$$max/min[0w_0 + 1w_1 + 2w_2 + 3w_3 + 4w_4], \qquad (7.3)$$

subject to the same constraints given above. We get $n_1(0)$ $(n_2(0))$ from the min (max) problems and the solutions are also in Table 7.3.

For $c = 1$ the alpha-cuts of $\overline{X}$ are obtained from equation (6.14) of Chapter 6. In general we may also get these alpha cuts from a linear programming problem. The expression pU in equation (6.14) is clearly an increasing function of p. Recall that $\overline{p} = (0.3/0.4/0.5)$. Let $\overline{X}[0] = [x_1(0), x_2(0)]$. The problems to solve are, a minimization problem

$$min[0.3U], \qquad (7.4)$$

using the smallest value of p, to get $x_1(0)$, and $U = \sum_{i=1}^{4} w_i$, subject to the constraints in equation (7.2), and the maximization problem

$$max[0.5U], \qquad (7.5)$$

using the largest value of p, to get $x_2(0)$, subject to the same constraints. Table 7.3 have the results.

Finally $\overline{R} = \frac{\overline{N}}{\overline{X}}$ whose α-cut equal to zero was calculated and put into Table 7.3. We used interval arithmetic to get the alpha zero cut of $\overline{R}$. Suppose $\overline{R}[0] = [r_1(0), r_2(0)]$. Then $[r_1(0), r_2(0)] = [n_1(0), n_2(0)]/[x_1(0), x_2(0)]$ which equals $[\frac{n_1(0)}{x_2(0)}, \frac{n_2(0)}{x_1(0)}]$.

The $\alpha = 1$ cuts in Table 7.3 were just doing the computations with real numbers.

In this example the system in very congested. From the $\alpha = 1$ cut in Table 7.3 we see that: (1) server utilization is almost 100%, which is high; (2) the average number of customers in the system is almost four, which is system capacity, and so we are probably loosing too many potential customers; and (3) average response time is almost 10 time units which is very long. We may consider increasing M, but only increasing system capacity will probably increase congestion and response time. So we consider the following: (1) adding another server; (2) getting a faster server which will increase $\overline{p}$; and (3) increasing system capacity. These type of considerations will be discussed in Chapters 13-17.

To see the variability in these answers we look at the α equal zero cut in Table 7.3. The $\alpha = 0$ cut is like a 99% confidence interval (see Chapter 3). We will measure the variability by the length of the $\alpha = 0$ cut. There is little variability in $\overline{U}$ $(\overline{N})$ because $\overline{U}$ $(\overline{N})$ is very close to its maximum value of one (four). Some variability shows up in $\overline{X}$ but the important fuzzy number $\overline{R}$ shows the greatest variability. The average response time can vary between 7.8 and 13.2. This variability is too great and the numbers are too large. We now want to look at ways to make $\overline{R}$ smaller. This will be discussed starting in Chapter 13.

7.3 References

1. Frontline Systems (www.frontsys.com).

2. D.A.Menasce and V.A.F.Almeida: Capacity Planning for Web Performance, Prentice Hall, Upper Saddle River, N.J., 1998.

3. X.Zheng, K.Reilly and J.J.Buckley: Applying Genetic Algorithms to Fuzzy Probability-Based Web Planning Models, Proceedings ACMSE, Savannah, Ga, 2003, pp. 241-245.

4. X.Zheng, K.Reilly and J.J.Buckley: Comparing Genetic Algorithms and Exhaustive Methods Used in Optimization Problems for Fuzzy Probability Based Web Planning Models, Proceedings 2003 Int. Conf. on AI, June 23-26, 2003, Las Vegas, Nevada. To appear.

Chapter 8

Computations: Two Servers

8.1 Introduction

Much of the calculations are the same here as in the one server case, so let us only discuss what is new for two servers. The crisp transition matrix is different and is shown in Table 8.1. In this table $q(i|s)$ denotes the probability that i customers leave the system during time period δ, given that s servers were busy at the start of the time period.

8.2 Calculations

Step 1

Let us briefly discuss finding $\overline{p}_{23}$ and $\overline{p}_{44}$. The fuzzy probabilities $\overline{q}(i|s)$ were computed in equations (3.38) and (3.39) of Chapter 3. If $\mathbf{S}_3$ is the statement "$q(i|2) \in \overline{q}(i|2)[\alpha]$, $i = 0, 1, 2$, $q(0|2) + q(1|2) + q(2|2) = 1$, $p(i) \in \overline{p}(i)[\alpha]$, $i = 0, 1, ..., 7$, $p(0) + ... + p(7) = 1$", then

$$\overline{p}_{23}[\alpha] = \{p(1)q(0|2) + p(2)q(1|2) + p(3)q(2|2)|\ \mathbf{S}_3\ \}. \tag{8.1}$$

Also

$$\overline{p}_{44}[\alpha] = \{p^*(1)q(1|2) + p^*(2)q(2|2) + q(0|2)|\ \mathbf{S}_3\ \}. \tag{8.2}$$

So, in general, we need mathematical software for the max/min of a non-linear function, subject to linear constraints (inequalities and equality) to be able to find the α-cuts of some of the fuzzy probabilities in the fuzzy transition matrix. This is done in the next chapter.

Future State

Previous State	0	1	2	3	4
0	$p(0)$	$p(1)$	$p(2)$	$p(3)$	$p^*(4)$
1	$p(0)q(1\|1)$	$p(1)q(1\|1)+$ $p(0)q(0\|1)$	$p(2)q(1\|1)+$ $p(1)q(0\|1))$	$p(3)q(1\|1)+$ $p(2)q(0\|1)$	$p^*(4)q(1\|1)+$ $p^*(3)q(0\|1)$
2	$p(0)q(2\|2)$	$p(0)q(1\|2)+$ $p(1)q(2\|2)$	$p(0)q(0\|2)+$ $p(1)q(1\|2)+$ $p(2)q(2\|2)$	$p(1)q(0\|2)+$ $p(2)q(1\|2)+$ $p(3)q(2\|2)$	$p^*(2)q(0\|2)+$ $p^*(3)q(1\|2)+$ $p^*(4)q(2\|2)$
3	0	$p(0)q(2\|2)$	$p(0)q(1\|2)+$ $p(1)q(2\|2)$	$p(0)q(0\|2)+$ $p(1)q(1\|2)+$ $p(2)q(2\|2)$	$p^*(1)q(0\|2)+$ $p^*(2)q(1\|2)+$ $p^*(3)q(2\|2)$
4	0	0	$p(0)q(2\|2)$	$p(0)q(1\|2)+$ $p(1)q(2\|2)$	$p^*(2)q(2\|2)+$ $p^*(1)q(1\|2)+$ $q(0\|2)$

Table 8.1: The Transition Matrix P for $c = 2$ and $M = 4$

Step 2

The same as for one server.

Step 3

U is server utilization, and with two servers, $U = \sum_{i=2}^{M} w_i$ so

$$\overline{U} = \sum_{i=2}^{M} \overline{w}_i, \tag{8.3}$$

which is evaluated by alpha cuts. In the crisp case $U \times 100$ is the percentage of time we expect **BOTH** servers to be busy.

$\overline{N}$, the fuzzy number of customers in the system, is calculated as the one server case.

$\overline{X}$ will now be different, having two servers. But $\overline{R}$ will still be $\overline{N}$ divided by $\overline{X}$. We will do the calculations for $c = 2$ and they can be easily extended to $c = 3$, or $C = 4$, etc.

We first derive the crisp expression for X and then fuzzify it. Let w_i, $1 \le i \le M$, be the crisp steady state probabilities and let p be the crisp probability that a customer leaves a server at the end of the time period δ, given that this customer was in the server at the start of the time period. Also define $L(i)$, $i = 0, 1, 2$, to be the probability that i customers leave a server at the end of the time period, with no conditions on how many servers were busy at the start of the time period. Then we see that

$$L(0) = w_0 + (1 - p)w_1 + (1 - p)^2 U_2, \tag{8.4}$$

where $U_2 = w_2 + \ldots + w_M$. Also

$$L(1) = pw_1 + 2p(1-p)U_2, \tag{8.5}$$

and

$$L(2) = p^2 U_2. \tag{8.6}$$

In the above equations the factors $(1-p)^2$, $2p(1-p)$ and p^2 come from the binomial probability distribution. In the binomial probability distribution $b(n, p)$, n is the number of independent experiments and p is the probability of a "success". Here $n = 2$ and a success is for a customer to leave a server. So, p^2 is the probability of two successes in two trials, $2p(1-p)$ is the probability of one success in two trials and $(1-p)^2$ is the probability of no successes in two trials. Then

$$X = 0L(0) + 1L(1) + 2L(2), \tag{8.7}$$

which simplifies to

$$X = pw_1 + 2pU_2. \tag{8.8}$$

Therefore, in the fuzzy case $\overline{X} = \overline{p}\,\overline{w}_1 + 2\overline{p}\overline{U}_2$ which is evaluated by α-cuts

$$\overline{X}[\alpha] = \{pw_1 + 2pU_2 | \ \mathbf{S} \ \}, \tag{8.9}$$

where $\mathbf{S}$ is "$p \in \overline{p}[\alpha]$, $w_i \in \overline{w}_i[\alpha]$, $1 \leq i \leq M$, $w_0 + \ldots + w_M = 1$" with $U_2 = w_2 + \ldots + w_M$.

Chapter 9

Example: Two Servers

9.1 Introduction

In this Chapter we will go through some of the computations discussed in Chapter 8. We will use the same data as in Chapter 7, except now $c = 2$, so we may compare the result of increasing the number of servers from $c = 1$ to $c = 2$. At the end of Chapter 7 we concluded that the system was too congested and we suggested adding another server. In this chapter we will see what happens when we do add another identical server.

So $c = 2$, $M = 4$, infinite calling source, the fuzzy probabilities of arrival are in Table 6.1 of Chapter 6. With two servers we need to use the fuzzy numbers $\overline{q}(i|s)$ for $i =$ zero to s and $s = 1, 2$. To compare to Chapter 7 we choose $\overline{p} = (0.3/0.4/0.5) = \overline{q}(1|1)$ and $1 - \overline{p} = \overline{q}(0|1)$. We now need to compute the α-cuts of $\overline{q}(i|2)$ from equation (3.39) in Chapter 3 for $i = 0, 1, 2$. We did this and the results are, for only $\alpha = 0$ and $\alpha = 1$: (1) $\overline{q}(0|2)[0] = [0.25, 0.49]$, $\overline{q}(1|2)[0] = [0.42, 0.50]$, $\overline{q}(2|2)[0] = [0.09, 0.25]$; and (2) $\overline{q}(0|2)[1] = 0.36$, $\overline{q}(1|2)[1] = 0.48$, $\overline{q}(2|2)[1] = 0.16$. The crisp transition matrix is in Table 8.1 of Chapter 8. We will only compute the $\alpha = 0$ and $\alpha = 1$ cut of the fuzzy steady state probabilities and the fuzzy numbers ($\overline{U}$, $\overline{R}$, etc.) for system performance.

9.2 Computations

Step 1

The first thing to do is to find the alpha equal zero cut of the fuzzy transition probabilities $\overline{p}_{ij}$ in the fuzzy transition matrix $\overline{P}$. These computations were discussed in Step 1 of Chapter 8. The intervals were put into interval matrix $I = (I_{ij})$. The result of these calculations are shown in Table 9.1.

I	Alpha Zero Cut	Alpha One Cut
I_{00}	$[0.07, 0.13]$	0.1
I_{01}	$[0.26, 0.34]$	0.3
I_{02}	$[0.17, 0.23]$	0.2
I_{03}	$[0.07, 0.13]$	0.1
I_{04}	$[0.2, 0.4]$	0.3
I_{10}	$[0.021, 0.065]$	0.04
I_{11}	$[0.127, 0.235]$	0.18
I_{12}	$[0.215, 0.307]$	0.26
I_{13}	$[0.12, 0.20]$	0.16
I_{14}	$[0.25, 0.47]$	0.36
I_{20}	$[0.0063, 0.0325]$	0.016
I_{21}	$[0.0528, 0.1500]$	0.096
I_{22}	$[0.1588, 0.2600]$	0.212
I_{23}	$[0.1675, 0.2749]$	0.22
I_{24}	$[0.3325, 0.5743]$	0.456
I_{30}	0	0
I_{31}	$[0.0063, 0.0325]$	0.016
I_{32}	$[0.0528, 0.1500]$	0.096
I_{33}	$[0.1588, 0.2600]$	0.212
I_{34}	$[0.5575, 0.7821]$	0.676
I_{40}	0	0
I_{41}	0	0
I_{42}	$[0.0063, 0.0325]$	0.016
I_{43}	$[0.0528, 0.1500]]$	0.096
I_{44}	$[0.8175, 0.9409]$	0.888

Table 9.1: Alpha Cuts of Fuzzy Transition Matrix, Two Servers

$\overline{w}$	Alpha Zero Cut	Alpha One Cut
$\overline{w}_0$	$[0.0001, 0.0048]$	0.0009
$\overline{w}_1$	$[0.0011, 0.0250]$	0.0064
$\overline{w}_2$	$[0.0111, 0.0788]$	0.0334
$\overline{w}_3$	$[0.0606, 0.1805]$	0.1138
$\overline{w}_4$	$[0.7176, 0.9265]$	0.8455

Table 9.2: Alpha Cuts of Fuzzy Steady State Probabilities, Two Servers

	Alpha Zero Cut	Alpha One Cut
$\overline{U}$	$[0.9702, 0.9988]$	0.9927
$\overline{N}$	$[3.5744, 3.9129]$	3.7966
$\overline{X}$	$[0.5896, 0.9994]$	0.7967
$\overline{R}$	$[3.5765, 6.6365]$	4.7653

Table 9.3: Alpha Cuts of $\overline{U}, \overline{N}, \overline{X}, \overline{R}$, Two Servers

Step 2

Now we use a genetic algorithm to estimate the end points of the alpha zero cut intervals for the fuzzy steady state probabilities $\overline{w}_i$, $0 \leq i \leq 4$. The results are in Table 9.2. We checked the results in Table 9.2 with the Premium Solver Platform from Frontline Systems [1].

Step 3

Finally, using $\overline{p}$ and the $\overline{w}_i$ we may find the $\alpha = 0$ and $\alpha = 1$ cuts of $\overline{U}, \overline{N}, \overline{X}$ and $\overline{R}$ whose fuzzy expressions were given in Step 3 of Chapter 8. The results are given in Table 9.3.

Let us now go through some of the details on how we got these α-cuts in Table 9.3. We compute the alpha-cuts of $\overline{U}$ from equation (8.3) of Chapter 8 by solving a linear programming problem. Let $\overline{w}_i[0] = [w_{i1}(0), w_{i2}(0)]$, $0 \leq i \leq 4$, which are obtained from Table 9.2. Let $\overline{U}[0] = [u_1(0), u_2(0)]$ The objective function is

$$max/min[w_2 + ... + w_4], \qquad (9.1)$$

subject to constraints

$$w_{i1}(0) \leq w_i \leq w_{i2}(0), i = 0, ..., 4, w_0 + ... + w_4 = 1. \qquad (9.2)$$

The result is in Table 9.3. The solution to the min (max) problem gives $u_1(0)$ $(u_2(0))$.

Alpha-cuts of $\overline{N}$ are given in equation (6.13) of Chapter 6. Let $\overline{N}[0] = [n_1(0), n_2(0)]$. Now we need to solve the following linear programming problem

$$max/min[0w_0 + 1w_1 + 2w_2 + 3w_3 + 4w_4], \qquad (9.3)$$

subject to the same constraints given above. We get $n_1(0)$ $(n_2(0))$ from the min (max) problems and the solutions are also in Table 9.3.

For $c = 2$ the alpha-cuts of $\overline{X}$ are obtained from equation (8.9) of Chapter 8. Let $\overline{X}[\alpha] = [x_1(\alpha), x_2(\alpha)]$. We want the $\alpha = 0$ cut. Clearly we obtain the end point $x_1(0)$ $(x_2(0))$ by using $p = 0.3$ $(p = 0.5)$ in equation (8.9). So

$$x_1(0) = min\{0.3w_1 + 0.6(w_2 + w_3 + w_4)| \mathbf{S}\}, \qquad (9.4)$$

and

$$x_2(0) = max\{0.5w_1 + 1.0(w_2 + w_3 + w_4)| \mathbf{S}\}, \qquad (9.5)$$

subject to the constraints in equation (9.2), where $\mathbf{S}$ was defined after equation (8.9) in Chapter 8. These are two linear programming problems which we solved using the simplex algorithm in Maple [2].

Finally $\overline{R} = \frac{\overline{N}}{\overline{X}}$ whose α-cut equal to zero was calculated and put into Table 9.3. We used interval arithmetic to get the alpha zero cut of $\overline{R}$. Suppose $\overline{R}[0] = [r_1(0), r_2(0)]$. Then $[r_1(0), r_2(0)] = [n_1(0), n_2(0)]/[x_1(0), x_2(0)]$ which equals $[\frac{n_1(0)}{x_2(0)}, \frac{n_2(0)}{x_1(0)}]$.

The $\alpha = 1$ cuts in Table 9.3 were just doing the computations with real numbers.

Now compare Tables 7.3 and 9.3. In both cases server utilization is high and the average number of customers in the system is almost four, which is system capacity. In both situations we are probably loosing too many potential customers. The major difference is in $\overline{X}$ and $\overline{R}$. $\overline{X}$ approximately doubled going from $c = 1$ to $c = 2$. The major change was in $\overline{R}$. The $\overline{R}$ for $c = 2$ is approximately half the $\overline{R}$ for $c = 1$. But also, the variability $\overline{R}$, as measured by its $\alpha = 0$ cut (like a 99% confidence interval) has been cut approximately in half by changing to $c = 2$ from $c = 1$.

9.3 References

1. Frontline Systems (www.frontsys.com).

2. Maple 6, Waterloo Maple Inc., Waterloo, Canada.

Chapter 10

Computations: Three or More Servers

In this chapter we will show that as c and M grow the computations in this model become prohibited, without a computer program to automatically do these calculations, and we will change to modeling a fuzzy queuing system using fuzzy arrival/sevice rates to be discussed in the next two chapters.

Let $c = 4$ and $M = 10$. Step 1 is to find the fuzzy probabilities in the fuzzy transition matrix. These will be the computations that become more and more involved as c and M grow larger. Given the fuzzy transition matrix, Step 2 is to determine the fuzzy steady state probabilities. In this case $\overline{P}$ will be 11×11 and we would use a genetic algorithm to estimate the end points of the intervals which would be the α-cuts of the fuzzy steady state probabilities $\overline{w}_i$, $0 \le i \le 10$. This computation, estimating alpha-cuts of the fuzzy steady state probabilities, is not too great for fuzzy transition matrices of size 20×20, or 50×50. Step 3 is getting alpha-cuts of the system parameters $\overline{U}, \overline{N}, \overline{R}$ and $\overline{X}$. The only one that can involve a growing computation, as c and M get larger, is $\overline{X}$ (see Step 3 in Chapter 7 and 9). However, determining $\overline{X}$ becomes minor compared to the problems of getting the fuzzy probabilities in the fuzzy transition matrix.

Now we will describe the $\overline{p}_{ij}$ in the fuzzy transition matrix $\overline{P}$ for $c = 4$ and $M = 10$. We first show the crisp P by listing its first five rows. We will employ the notation $i - j - k$ for $p(i)q(j|k)$ and i^* is for $p^*(i) = \sum_{k=i}^{\infty} p(k)$. We number the rows and columns of P as $0, 1, 2..., 10$. The first five rows are:

1. row #0: $p(0), ..., p^*(10)$.

2. row #1: $0 - 1 - 1, (0 - 0 - 1) + (1 - 1 - 1), (1 - 0 - 1) + (2 - 1 - 1), (2 - 0 - 1) + (3 - 1 - 1), ..., (9^* - 0 - 1) + (10^* - 1 - 1)$.

3. row #2: $0 - 2 - 2, (0 - 1 - 2) + (1 - 2 - 2), (0 - 0 - 2) + (1 - 1 - 2) + (2 - 2 - 2), (1 - 0 - 2) + (2 - 1 - 2) + (2 - 2 - 2), ..., (8^* - 0 - 2) + (9^* - $

$1 - 2) + (10^* - 2 - 2)$.

4. row#3: $0 - 3 - 3, (0 - 2 - 3) + (1 - 3 - 3), (0 - 1 - 3) + (1 - 2 - 3) + (2 - 3 - 3), ((0 - 0 - 3) + (1 - 1 - 3) + (2 - 2 - 3) + (3 - 3 - 3), ..., (7^* - 0 - 3) + (8^* - 1 - 3) + (9^* - 2 - 3) + (10^* - 3 - 3)$.

5. row #4: $0 - 4 - 4, (0 - 3 - 4) + (1 - 4 - 4), (0 - 2 - 4) + (1 - 3 - 4) + (2 - 4 - 4), (0 - 1 - 4) + (1 - 2 - 4) + (2 - 3 - 4) + (3 - 4 - 4), (0 - 0 - 4) + (1 - 1 - 4) + (2 - 2 - 4) + (3 - 3 - 4) + (4 - 4 - 4), ..., (6^* - 0 - 4) + (7^* - 1 - 4) + (8^* - 2 - 4) + (9^* - 3 - 4) + (10^* - 4 - 4)$.

There is some simplification in rows #6 through #10 because $p_{50} = 0$, $p_{60} = p_{61} = 0, ..., p_{10,0} = ... = p_{10,5} = 0$.

So as c grows to 10, 20,... and M increases from 50, 100, etc., the description of crisp P become more complicated. There is a definite pattern in P from Table 6.2 to Table 8.1 to the list above, but the row descriptions still become more complicated.

Using crisp P and the values of $\overline{p}(i)$, $0 \leq i \leq L$, where $\overline{p}(i) = 0$ for $i > L$, and the values of $\overline{q}(j|k)$, $j = 0, 1, 2, ..., k$, $k = 1, 2, 3, ..., c$ we still need to get α-cuts of the fuzzy probabilities $\overline{p}_{ij}$ in $\overline{P}$. Many of these calculations require a non-linear optimizer like Premium Solver Platform V5.0 from Frontline Systems [1]. For example, consider $\overline{p}_{44}$ whose alpha-cuts are solutions to the non-linear optimization problem

$$max/min[p(0)q(0|4) + p(1)q(1|4) + p(2)q(2|4) + p(3)q(3|4) + p(4)q(4|4)],$$
$$(10.1)$$

subject to the constraints

$$p(i) \in \overline{p}(i)[\alpha], 0 \leq i \leq L, p(0) + ... + p(L) = 1, \tag{10.2}$$

and

$$q(i|4) \in \overline{q}(i|4)[\alpha], 0 \leq i \leq 4, q(0|4) + ... + q(4|4) = 1. \tag{10.3}$$

Because of the excessive computations needed in this model of the system, using fuzzy probabilities for $c \geq 4$ and $M \geq 10$, without a computer program to do these calculations for us, let us consider another method of modeling a fuzzy queuing system in the next two chapters.

10.1   References

1. Frontline Systems (www.frontsys.com).

Chapter 11

Fuzzy Arrival/Service Rates

11.1 Introduction

In this chapter we will model the system using the arrival rate λ and the service rate μ for any server. This is a common method used in queuing theory ([1],[2]). We first discuss the crisp model. The system has c parallel and identical servers, system capacity M (in the servers and in the queue) and an infinite calling source. The λ rate will be state independent, which means that λ does not depend on how many customers are in the system. But if there are n customers in the system, then the rate of departure from the whole system is $\mu_n = n\mu$, for $1 \leq n < c$ and $\mu_n = c\mu$ for $c \leq n \leq M$.

A basic assumption is that we are in steady-state, all transient behavior has died down and can be neglected, and the time interval δ is sufficiently small so that the probability of two or more events occurring during δ is zero. If we are in state n, or there are n customers in the system with $0 < n < M$, we can have only two events occurring: (1) one customer arrives and we have $n + 1$ in the system; or (2) one customer finishes service and leaves and we have $n - 1$ left in the system. Usually for steady state we assume that $\lambda \leq \mu$ when we have infinite capacity. However, since we have finite system capacity we do not need to assume that $\lambda \leq \mu$. If we are in state zero $(n = 0)$, we can only go to $n = 1$ and we can get to state $n = 0$ from $n = 1$ when a customer leaves service. We can get to state $n = M$ only from $n = M - 1$ with an arrival and we can leave state $n = M$ to state $M - 1$ when a customer leaves service.

The main objective is to compute the steady state probabilities w_i, $0 \leq i \leq M$, from which we may determine various measures of system performance.

Using a transition rate diagram, the expected rate of flow into state n is

$$\lambda w_{n-1} + \mu_{n+1} w_{n+1}, \tag{11.1}$$

and the expected rate of flow out of state n is

$$\lambda w_n + \mu_n w_n, \tag{11.2}$$

for $0 < n < M$. We set these two equal to get the balance equation

$$\lambda w_{n-1} + \mu_{n+1} w_{n+1} = \lambda w_n + \mu_n w_n, \tag{11.3}$$

for $0 < n < M$. The balance equation for state $n = 0$ is simply $\lambda w_0 = \mu_1 w_1$ and for $n = M$ it is $\lambda w_{M-1} = \mu_M w_M$. We solve these balance equations for w_i, $1 \leq i \leq M$, functions of w_0 and then use the fact that the sum of all the w_i must equal one to obtain a formula for w_0. The final result is that $w_i = F_i(\lambda, \mu, c, M)$, $0 \leq i \leq M$. That is, the steady state probabilities are function of λ, μ, c and M ([1],[2]).

Now we can determine U, N, X and R, all defined in Chapter 6. U, if $c = 1$, equation (6.9), N, equation (6.12) and $R = \frac{N}{X}$ are all straight forward. X is the expected number of customers leaving the system per time period δ. So ([1], Chapter 8)

$$X = \sum_{k=1}^{c-1} \mu_k w_k + c\mu \sum_{k=c}^{M} w_k. \tag{11.4}$$

11.2   Fuzzy Steady State Probabilities

The parameters $\lambda =$ the expected (average) number of arrivals per time unit δ, and $\mu =$ the expected (average) number of service completions per time unit δ, are usually unknown and must be estimated. As discussed in Section 3.3 of Chapter 3, we use confidence interval to obtain fuzzy numbers $\overline{\lambda}$ for λ and $\overline{\mu}$ for μ. Now we want to calculate the fuzzy steady state probabilities.

From the preceding section we have $w_i = F_i(\lambda, \mu, c, M)$. Then

$$\overline{w}_i = F_i(\overline{\lambda}, \overline{\mu}, c, M), \tag{11.5}$$

to be evaluate using α-cuts (equations (2.24) and (2.25) of Chapter 2). Then we have

$$\overline{w}_i[\alpha] = \{F_i(\lambda, \mu, c, M) \mid \lambda \in \overline{\lambda}[\alpha], \mu \in \overline{\mu}[\alpha]\}, \tag{11.6}$$

for all α in $[0, 1]$. Expressions for F_i can be looked up in [1], or [2], and then we can find α-cuts of the fuzzy steady state probabilities. We will do this in an example in the following chapter.

Let us explain why equation (11.6) will produce the correct α-cuts for the fuzzy steady state probabilities. Choose and fix a value of α in [0, 1]. Next choose a value of λ in $\overline{\lambda}[\alpha]$ and a value of μ in $\overline{\mu}[\alpha]$. Using this arrival rate λ and this service rate μ we use the balance equations presented above to compute w_i, $0 \leq i \leq M$. We get $w_i = F_i(\lambda, \mu, c, M)$, $i = 0, ..., M$. Now do this for all choices of λ in $\overline{\lambda}[\alpha]$ and all choices of μ in $\overline{\mu}[\alpha]$. What we get is equation (11.6).

11.3 Fuzzy System Parameters

Computing $\overline{U}$ and $\overline{N}$ from the fuzzy steady state probabilities is the same as was done in Chapters 6-9. Also, $\overline{R} = \frac{\overline{N}}{\overline{X}}$ as in Chapters 6-9. So let us look at $\overline{X}$. We simply fuzzify the crisp equation for X given above. The justification is the same as given for equation (11.6). Therefore, the α-cuts of $\overline{X}$ are

$$\overline{X}[\alpha] = \{\sum_{k=1}^{c-1} \mu_k w_k + c \sum_{k=c}^{M} \mu w_k | \quad \mathbf{S} \quad \}, \tag{11.7}$$

where the statement $\mathbf{S}$ is "$w_i \in \overline{w}_i[\alpha]$, $0 \leq i \leq M$, $w_0 + ... + w_M = 1$, $\mu \in \overline{\mu}[\alpha]$". In particular, when $c = 1$ we have

$$\overline{X}[\alpha] = \{\mu(w_1 + ... + w_M)| \quad \mathbf{S} \quad \}, \tag{11.8}$$

and if $c = 2$

$$\overline{X}[\alpha] = \{\mu w_1 + 2\mu(w_2 + ... + w_M)| \quad \mathbf{S} \quad \}, \tag{11.9}$$

with $\mathbf{S}$ defined above.

In the next chapter we will obtain numerical results for these calculations.

11.4 References

1. D.A.Menasce and V.A.F.Almeida: Capacity Planning for Web Performance, Prentice Hall, Upper Saddle River, N.J., 1998.

2. H.A.Taha: Operations Research, Fifth Edition, Macmillan, N.Y., 1992.

Chapter 12

Example: Fuzzy Arrival/Service Rates

12.1 Introduction

In this chapter we will look at two examples. The values of the parameters c and M are chosen to match those used in the two examples discussed in Chapters 7 and 9. We can not directly compare the numerical results $(\overline{U}, \overline{N}, \overline{X}, \overline{R})$ between those in Chapters 7 and 9 to those in this chapter because the two models are very different. In Chapters 7 and 9 we started with fuzzy probabilities, computed a fuzzy transition matrix and then the fuzzy steady state probabilities before we obtained the fuzzy numbers for $\overline{U}, \overline{N}, \overline{X}, \overline{R}$. In this chapter we start with fuzzy numbers for the arrival rate and the service rate, then find the fuzzy steady state probabilities and the fuzzy numbers for $\overline{U}, \overline{N}, \overline{X}, \overline{R}$. In the model based on fuzzy probabilities we do not relate the fuzzy steady state probabilities, fuzzy arrival rate and the fuzzy service rate as in the balance equations (equation (11.3), Chapter 11). The expressions for $\overline{U}, \overline{N}, \overline{R}$ are the same in both models, $\overline{X}$ is computed differently in the two models. What we can compare is the computations needed in the model presented in Chapter 11 and this chapter to those seen in Chapters 7 and 9.

12.2 One Server

Let $c = 1$ and $M = 4$ as in Chapter 7. Let the fuzzy arrival rate be $\overline{\lambda} = (3/4/5)$ and the fuzzy service rate be $\overline{\mu} = (5/6/7)$. These are both triangular fuzzy numbers. Obtaining fuzzy numbers for λ and μ, through the use of confidence intervals, was discussed in Section 3.3 of Chapter 3.

Step 1

The first thing to do is calculate the fuzzy steady state probabilities $\overline{w}_i$, $0 \leq i \leq 4$. The crisp expressions are ([3],[4])

$$w_n = \rho^n w_0, \tag{12.1}$$

for $0 \leq n \leq 4$, where $\rho = \lambda/\mu$, and

$$w_0 = \left\{ \begin{array}{ll} \frac{1-\rho}{1-\rho^5}, & \rho \neq 1 \\ \frac{1}{5}, & \rho = 1. \end{array} \right. \tag{12.2}$$

As explained in the previous chapter all we need to do is to fuzzify equations (12.1) and (12.2). We calculate the fuzzy steady state probabilities by their α-cuts. So

$$\overline{w}_n[\alpha] = \{\rho^n \frac{1-\rho}{1-\rho^5}| \quad \mathbf{S} \quad \}, \tag{12.3}$$

for $n = 0, 1, 2, 3, 4$, assuming $\rho \neq 1$, and the statement $\mathbf{S}$ is "$\lambda \in \overline{\lambda}[\alpha]$, $\mu \in \overline{\mu}[\alpha]$", for all α in $[0, 1]$.

As in Chapter 7 we will only compute the $\alpha = 0$ and $\alpha = 1$ cuts. The results are given in Table 12.1. We will go through the details of finding the $\alpha = 0$ cut for $\overline{w}_2$ in the next example.

Example 12.2.1

Let $\overline{w}_2[0] = [w_{21}, w_{22}]$. Let $z = f(\lambda, \mu) = \rho^2 w_0$, $\rho = \lambda/\mu$, the crisp value of w_2, with w_0 given in equation (12.2). Also let $\overline{\lambda}[0] = [\lambda_1, \lambda_2]$ and $\overline{\mu}[0] = [\mu_1, \mu_2]$. The optimization problems to solve to get the end points of the alpha zero cut of $\overline{w}_2$ are

$$w_{21} = min \, f(\lambda, \mu), \tag{12.4}$$

and

$$w_{22} = max \, f(\lambda, \mu), \tag{12.5}$$

subject to the constraints

$$\lambda_1 \leq \lambda \leq \lambda_2, \ \mu_1 \leq \mu \leq \mu_2. \tag{12.6}$$

This is a non-linear optimization problem subject to linear inequality constraints.

However, since there are only two variables λ and μ we may construct the graph of $z = f(\lambda, \mu)$ and sometimes easily see from this surface the solution for w_{21} and w_{22}. Now $\overline{\lambda} = (3/4/5)$ and $\overline{\mu} = (5/6/7)$ so the constraints are $3 \leq \lambda \leq 5$ and $5 \leq \mu \leq 7$. So we need to look at the surface over the square $[3, 5] \times [5, 7]$ and find the highest point being w_{22} and the lowest point which is w_{21}. The graph of $z = f(\lambda, \mu)$ over the square is shown in Figure 12.1

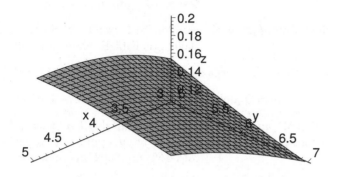

Figure 12.1: Fuzzy Steady State Probability in Example 12.2.1

where $x = \lambda$ and $y = \mu$. We can not use $\lambda = \mu = 5$ since we will get $0/0$ in $x = f(\lambda, \mu)$ so the graph is for $3 \le \lambda \le 4.9999$ and $5 \le \mu \le 7$. From this figure we see that $w_{21} = 0.1065$ at $\lambda = 3$ and $\mu = 7$; and $w_{22} = 0.2000$ for $\lambda = 5$ and $\mu = 5$, $f(5,7) = 0.1791$. Notice that when $\lambda = \mu = 5$ we have $w_i = 0.2000$ for all i.

We will need to solve these non-linear optimization problems hundreds of times in Chapters 13-17 and we can not be studying hundreds of surfaces. So we will usually use a non-linear optimization package to do the job. We downloaded the Premium Solver Platform V5.0 from Frontline Systems [1] as an add-on to the add-in "Solver" in Microsoft Excel. It has a non-linear optimizer to find maximums and minimums subject to equality and inequality constraints. However, it can get "stuck" in a local minimum (or maximum) so use the "multiple start" feature so that the algorithm can randomly choose many starting points in the feasible space to obtain the true minimum (maximum). Another problem is when $\lambda = \mu$ we have $0/0$ and the optimizer can fail. We know the correct value of $f(\lambda, \mu)$ in all these cases (0.2000 if $M = 4$ and $1/11$ for $M = 10$ in future chapters) and then we go back to the surfaces, as in Example 12.2.1, to obtain the solution.

$\overline{w}$	Alpha Zero Cut	Alpha One Cut
$\overline{w}_0$	$[0.2000, 0.5798]$	0.3839
$\overline{w}_1$	$[0.2000, 0.2608]$	0.2559
$\overline{w}_2$	$[0.1065, 0.2000]$	0.1706
$\overline{w}_3$	$[0.0456, 0.2000]$	0.1137
$\overline{w}_4$	$[0.0196, 0.2000]$	0.0758

Table 12.1: Alpha Cuts of the Fuzzy Steady State Probabilities, One Server

Step 2

Using the $\alpha = 0$ and $\alpha = 1$ cuts of the $\overline{w}_n$ we may find the $\alpha = 0$ and $\alpha = 1$ cuts of $\overline{U}, \overline{N}, \overline{X}$ and $\overline{R}$ whose fuzzy expressions were presented in Chapter 7. The results are given in Table 12.2. The $\alpha = 0$ cut is like a 99% confidence interval.

Let us now go through some of the details on how we got these α-cuts in Table 12.2. We compute the alpha-cuts of $\overline{U}$ from equations (7.1) and (7.2) in Chapter 7 by solving a linear programming problem. Let $\overline{w}_n[0] = [w_{n1}(0), w_{n2}(0)]$, $0 \le n \le 4$, which are obtained from Table 12.1. Let $\overline{U}[0] = [u_1(0), u_2(0)]$ The objective function is

$$max/min[w_1 + ... + w_4],\qquad(12.7)$$

subject to constraints

$$w_{i1}(0) \le w_i \le w_{i2}(0), i = 0, ..., 4, w_0 + ... + w_4 = 1.\qquad(12.8)$$

The result is in Table 12.2. The solution to the min (max) problem gives $u_1(0)$ $(u_2(0))$.

Alpha-cuts of $\overline{N}$ are given in equation (7.3) in Chapter 7. Let $\overline{N}[0] = [n_1(0), n_2(0)]$. Now we need to solve the following linear programming problem

$$max/min[0w_0 + 1w_1 + 2w_2 + 3w_3 + 4w_4],\qquad(12.9)$$

subject to the same constraints given above. We get $n_1(0)$ $(n_2(0))$ from the min (max) problems and the solutions are also in Table 12.2.

The α-cuts of $\overline{X}$, from equation (11.7) in Chapter 11, are

$$\overline{X}[\alpha] = \{\sum_{n=1}^{4} \mu w_n | \quad \mathbf{S} \quad \},\qquad(12.10)$$

where the statement $\mathbf{S}$ is "$w_n \in \overline{w}_n[\alpha]$, $0 \le n \le 4$, $w_0 + ... + w_4 = 1$, $\mu \in \overline{\mu}[\alpha]$". We want the $\alpha = 0$ cut for Table 12.2. Notice that the expression to be evaluated in equation (12.10) is an increasing function of μ. So we

	Alpha Zero Cut	Alpha One Cut
$\overline{U}$	$[0.4202, 0.8000]$	0.6161
$\overline{N}$	$[0.6767, 2.000]$	1.2424
$\overline{X}$	$[2.101, 5.600]$	3.6966
$\overline{R}$	$[0.1208, 0.9519]$	0.3358

Table 12.2: Alpha Cuts of $\overline{U}, \overline{N}, \overline{X}, \overline{R}$, One Server

can find the end points of the $\alpha = 0$ cut by solving two linear programming problems. For the left end point we want

$$min[5(w_1 + ... + w_4)], \qquad (12.11)$$

using the smallest value of μ in $\overline{\mu}[0] = [5, 7]$, and for the right end point we need to get

$$max[7(w_1 + ... + w_4)], \qquad (12.12)$$

using the largest value on μ. Use the same constraints. The solution is in Table 12.2.

Finally $\overline{R} = \frac{\overline{N}}{\overline{X}}$ whose α-cut equal to zero was calculated and put into Table 12.2. We used interval arithmetic to get the alpha zero cut of $\overline{R}$. Suppose $\overline{R}[0] = [r_1(0), r_2(0)]$. Then $[r_1(0), r_2(0)] = [n_1(0), n_2(0)]/[x_1(0), x_2(0)]$ which equals $[\frac{n_1(0)}{x_2(0)}, \frac{n_2(0)}{x_1(0)}]$.

The $\alpha = 1$ cuts in Table 12.2 were just doing the computations with real numbers, $\lambda = 4$ and $\mu = 6$.

All the linear programming problems will be solved using "simplex" in Maple [2].

12.3 Two Servers

Let $c = 2$ and $M = 4$ as in Chapter 9. Let the fuzzy arrival rate be $\overline{\lambda} = (3/4/5)$ and the fuzzy service rate be $\overline{\mu} = (5/6/7)$ as in the previous section. Obtaining fuzzy numbers, through the use of confidence interval, for λ (μ) was discussed in Section 3.3 of Chapter 3.

Step 1

We first need to calculate the fuzzy steady state probabilities $\overline{w}_i$, $0 \leq i \leq 4$. The crisp expressions are ([3],[4]): (1)$w_1 = \rho w_0$; (2) $w_2 = (\rho^2/2)w_0$; (3) $w_3 = (\rho^3/4)w_0$; and (4) $w_4 = (\rho^4/8)w_0$ where $\rho = \lambda/\mu$ and

$$w_0 = [1 + \rho + \frac{\rho^2(1 - \{\rho^3/8\})}{2(1 - \rho/2)}]^{-1}, \qquad (12.13)$$

$\overline{w}$	Alpha Zero Cut	Alpha One Cut
$\overline{w}_0$	$[0.3478, 0.6475]$	0.5031
$\overline{w}_1$	$[0.2775, 0.3503]$	0.3354
$\overline{w}_2$	$[0.0595, 0.1739]$	0.1118
$\overline{w}_3$	$[0.0127, 0.0870]$	0.0373
$\overline{w}_4$	$[0.0027, 0.0435]$	0.0124

Table 12.3: Alpha Cuts of the Fuzzy Steady State Probabilities, Two Servers

if $\rho \neq 2$ and

$$w_0 = [1 + \rho + 1.5\rho^2]^{-1}, \tag{12.14}$$

when $\rho = 2$. As explained in the previous chapter all we need to do is to fuzzify these equations for the crisp w_i, $0 \le i \le 4$. We calculate the fuzzy steady state probabilities by their α-cuts. So

$$\overline{w}_1[\alpha] = \{\rho w_0 | \quad \mathbf{S} \quad \}, \tag{12.15}$$

$$\overline{w}_2[\alpha] = \{\rho^2 w_0/2 | \quad \mathbf{S} \quad \}, \tag{12.16}$$

$$\overline{w}_2[\alpha] = \{\rho^3 w_0/4 | \quad \mathbf{S} \quad \}, \tag{12.17}$$

$$\overline{w}_4[\alpha] = \{\rho^4 w_0/8 | \quad \mathbf{S} \quad \}, \tag{12.18}$$

where $(\rho \neq 2)$

$$\overline{w}_0[\alpha] = \{\, expression\, for\, w_0 \mid \quad \mathbf{S} \quad \}, \tag{12.19}$$

and the statement $\mathbf{S}$ is "$\lambda \in \overline{\lambda}[\alpha]$, $\mu \in \overline{\mu}[\alpha]$", for all α in $[0,1]$.

As in Chapter 9 we will only compute the $\alpha = 0$ and $\alpha = 1$ cuts. The results are given in Table 12.3. We consider the details of determining $\overline{w}_0[0]$ in the next example.

Example 12.3.1

This example just extends Example 12.2.1 to the expression given in equation (12.13) for w_0. Substitute x for λ and y for μ and then $\rho = x/y$ into equation (12.13) and we obtain z a function of x and y for (x, y) in the square $[3,5] \times [5,7]$. The graph of this surface is in Figure 12.2. If $\overline{w}_0[0] = [w_{01}, w_{02}]$, we see from Figure 12.2 that the highest point is $w_{02} = 0.6475$ at $x = 3$ and $y = 7$ and the lowest point on the surface is $w_{01} = 0.3478$ at $x = y = 5$.

Step 2

Using the $\alpha = 0$ and $\alpha = 1$ cuts of the $\overline{w}_n$ we may find the $\alpha = 0$ and $\alpha = 1$ cuts of $\overline{U}, \overline{N}, \overline{X}$ and $\overline{R}$ whose fuzzy expressions were presented in Section 11.3

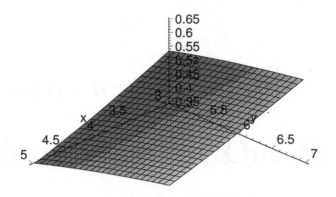

Figure 12.2: Fuzzy Steady State Probability in Example 12.3.1

of Chapter 11. The results are given in Table 12.4. The base of these fuzzy numbers, the alpha zero cut, is similar to a 99% confidence interval.

Let us now go through some of the details on how we got these α-cuts in Table 12.4. We compute the alpha-cuts of $\overline{U}$, from equations (9.1) and (9.2) in Chapter 9, as a solution to a linear programming problem. Let $\overline{w}_n[0] = [w_{n1}(0), w_{n2}(0)]$, $0 \le n \le 4$, which are obtained from Table 12.3. Let $\overline{U}[0] = [u_1(0), u_2(0)]$. The objective function is

$$max/min[w_2 + ... + w_4], \tag{12.20}$$

subject to constraints

$$w_{i1}(0) \le w_i \le w_{i2}(0), i = 0, ..., 4, w_0 + ... + w_4 = 1. \tag{12.21}$$

The result is in Table 12.4. The solution to the min (max) problem gives $u_1(0)$ $(u_2(0))$.

Alpha-cuts of $\overline{N}$ are given in equation (7.3) in Chapter 7. Let $\overline{N}[0] = [n_1(0), n_2(0)]$. Now we need to solve the following linear programming problem

$$max/min[0w_0 + 1w_1 + 2w_2 + 3w_3 + 4w_4], \tag{12.22}$$

subject to the same constraints given above. We get $n_1(0)$ $(n_2(0))$ from the min (max) problems and the solutions are also in Table 12.4.

The α-cuts of $\overline{X}$, from equation (11.7) in Chapter 11, are

$$\overline{X}[\alpha] = \{\mu w_1 + 2\mu w_2 + 2\mu w_3 + 2\mu w_4| \quad \mathbf{S} \quad \}, \tag{12.23}$$

where the statement $\mathbf{S}$ is "$w_n \in \overline{w}_n[\alpha]$, $0 \le n \le 4$, $w_0 + ... + w_4 = 1$, $\mu \in \overline{\mu}[\alpha]$". We want the $\alpha = 0$ cut for Table 12.4. Notice that the expression to be evaluated in equation (12.23) is an increasing function of μ. So we

	Alpha Zero Cut	Alpha One Cut
$\overline{U}$	$[0.0749, 0.3044]$	0.1615
$\overline{N}$	$[0.4455, 1.1306]$	0.7205
$\overline{X}$	$[2.137, 6.6962]$	3.9504
$\overline{R}$	$[0.0665, 0.5291]$	0.1824

Table 12.4: Alpha Cuts of $\overline{U},\overline{N},\overline{X},\overline{R}$, Two Servers

can find the end points of the $\alpha = 0$ cut by solving two linear programming problems. For the left end point we want

$$min[5w_1 + 10(w_2 + ... + w_4)], \qquad (12.24)$$

using the smallest value of μ in $\overline{\mu}[0] = [5, 7]$, and for the right end point we need to get

$$max[7w_1 + 14(w_2 + ... + w_4)], \qquad (12.25)$$

using the largest value on μ. Use the same constraints. The solution is in Table 12.4.

Finally $\overline{R} = \frac{\overline{N}}{\overline{X}}$ whose α-cut equal to zero was calculated and put into Table 12.2. We used interval arithmetic to get the alpha zero cut of $\overline{R}$. Suppose $\overline{R}[0] = [r_1(0), r_2(0)]$. Then $[r_1(0), r_2(0)] = [n_1(0), n_2(0)]/[x_1(0), x_2(0)]$ which equals $[\frac{n_1(0)}{x_2(0)}, \frac{n_2(0)}{x_1(0)}]$.

The $\alpha = 1$ cuts in Table 12.4 were just doing the computations with real numbers, $\lambda = 4$ and $\mu = 6$.

All the linear programming problems were solved using "simplex" in Maple [2].

12.4   Three or More Servers

As in Chapter 10 let us assume that $c = 4$ and $M = 10$. The equations for the crisp steady state probabilities are just extensions of those given in Section 12.3. We fuzzify these crisp equations to get $\overline{w}_i$, $0 \le i \le 10$. Then

$$\overline{U} = \sum_{i=4}^{10} \overline{w}_i, \qquad (12.26)$$

whose alpha-cuts are found by solving a linear programming problem. $\overline{N}$ and $\overline{R}$ are computed the same way for all c. To obtain $\overline{X}$, from equation (11.7) in Chapter 11, let

$$H(\mu, w) = \mu(w_1 + w_2 + w_3) + 4\mu(w_4 + ... + w_{10}). \qquad (12.27)$$

then
$$\overline{X}[\alpha] = \{H(\mu, w) | \, \mathbf{S}\,\},\qquad\qquad(12.28)$$

for all α where $\mathbf{S}$ is "$\mu \in \overline{\mu}[\alpha]$, $w_i \in \overline{w}_i[\alpha]$, $0 \le i \le 10$, $w_0 + ... + w_{10} = 1$".

12.5 References

1. Frontline Systems (www.frontsys.com).

2. Maple 6, Waterloo Maple Inc., Waterloo, Canada.

3. D.A.Menasce and V.A.F.Almeida: Capacity Planning for Web Performance, Prentice Hall, Upper Saddle River, N.J., 1998.

4. H.A.Taha: Operations Research, Fifth Edition, Macmillan, N.Y., 1992.

Chapter 13

Optimization: Without Revenue/Costs

13.1 Introduction

We have two models for the fuzzy queuing system: (1) using fuzzy probabilities in section 13.2; and (2) using fuzzy arrival/service rates in section 13.3. Sections 13.2 and 13.3 are written so they can be read independently, so the model details in 13.2 are repeated in 13.3.

13.2 Fuzzy Probabilities

Suppose we decide that we would like to increase $\overline{U}$, server utilization, and also decrease $\overline{R}$, average response time. In other words, we want to find the values of the variables to maximize $\overline{U}$ and/or minimize $\overline{R}$. The variables are $M = 4, 5, ..., 10$, $c = 1, 2, 3$ and two possible values for $\overline{p}$. Assume that the $\overline{p}(i)$, the fuzzy probabilities for arrivals, have be computed and they are fixed. The two values for $\overline{p}$ are $\overline{p}_1$ and $\overline{p}_2$, which give the fuzzy probability that a customer leaves a server in time interval δ, given the server was busy at the start of the time interval. $\overline{p}_1$ is for the servers we are now using and the new $\overline{p}_2$ is for faster servers we are considering using. So both values of $\overline{p}$ are in $[0, 1]$ but $\overline{p}_2$ is $\overline{p}_1$ shifted to the right, for a faster server. We obtained $\overline{p}_1$ from data, as discussed in Chapter 3, but having no experience with the new server $\overline{p}_2$ was determined through "expert" opinion (Section 3.4 of Chapter 3). We assume that all servers are the same which means they all have property $\overline{p}_1$ or they all have $\overline{p}_2$.

We will now present two methods of optimization. The first one is to minimize the fuzzy set $\overline{R}$, or maximize $\overline{U}$, and the second one ranks the fuzzy sets from smallest to largest.

13.2.1   Minimize $\overline{R}$

We will first consider minimizing $\overline{R}$, a similar procedure can be employed for maximizing $\overline{U}$, and then we will discuss optimizing both.

Now $\overline{R}$ is a function of c, M and $\overline{p}$ and we show this as writing $\overline{Z} = \overline{R}(M, c, \overline{p})$. So, for different values of the variables we get triangular shaped fuzzy numbers $\overline{Z}$. Find the values of the variables to minimize $\overline{Z}$

We discussed the problem of minimizing a fuzzy set in Section 2.5 of Chapter 2, so we will only give a brief discussion here. For our problem let: (1) m_z be the center of the core of $\overline{Z}$ (the core of a fuzzy number is the interval where the membership function equals one); (2) L_z be the area under the graph of the membership function to the left of m_z; and (3) R_z be the area under the graph of the membership function to the right of m_z. See Figure 2.5. In our application the core of $\overline{Z}$ will be a single point. For $min\overline{R}$ we substitute: (1) $min[m_z]$; (2) $maxL_z$, or maximize the possibility of obtaining values less than m_z; and (3) $minR_z$, or minimize the possibility of obtaining values greater than m_z. So for $min\overline{R}$ we have

$$V = (maxL_z, min[m_z], minR_z). \tag{13.1}$$

First let K_1 be a sufficiently large positive number so that $maxL_z$ is equivalent to $minL_z^*$ where $L_z^* = K_1 - L_z$. The multiobjective problem is

$$minV = (minL_z^*, min[m_z], minR_z). \tag{13.2}$$

One way to explore the undominated set (see Section 2.5) is to change the multiobjective problem into a single objective. The single objective problem is

$$min(\lambda_1[K_1 - L_z] + \lambda_2 m_z + \lambda_3 R_z), \tag{13.3}$$

where $\lambda_i > 0$, $1 \leq i \leq 3$, $\lambda_1 + \lambda_2 + \lambda_3 = 1$, and the values of the variables are

$$c = 1, 2, 3; \quad M = 4, 5, ..., 10; \quad \overline{p} = \overline{p}_i, i = 1, 2. \tag{13.4}$$

You will get different undominated solutions by choosing different values of $\lambda_i > 0$, $\lambda_1 + \lambda_2 + \lambda_3 = 1$. The decision maker is to choose the values of the weights λ_i for the three minimization goals. Usually one picks different values for the λ_i to explore the solution set and then lets the decision maker choose an optimal solution from this set of solutions. It is usually best to present management with a number of optimal solutions, instead of only one optimal solution. Managers are decision makers, and when they have multiple solutions to pick from, they can decide on one of them weighing the various alternatives associated with each.

This is how we propose to handle the problem of $min\overline{R}$ in finding the optimal system. Numerical solutions to this optimization problem for continuous variables can be difficult. In the past we have employed an evolutionary algorithm to generate good approximate solutions in the continuous case. However, here we are in a discrete situation ($c = 1, 2, M = 4, 10$,etc.).

Case	Number of Servers	System Capacity	Server Rate
1	$c = 1$	$M = 4$	$\overline{p}_1$
2	$c = 1$	$M = 10$	$\overline{p}_1$
3	$c = 2$	$M = 4$	$\overline{p}_1$
4	$c = 2$	$M = 10$	$\overline{p}_1$
5	$c = 1$	$M = 4$	$\overline{p}_2$
6	$c = 1$	$M = 10$	$\overline{p}_2$
7	$c = 2$	$M = 4$	$\overline{p}_2$
8	$c = 2$	$M = 10$	$\overline{p}_2$

Table 13.1: The Eight Cases in Example 13.2.1.1

Example 13.2.1.1

In this example we will use the $\overline{p}(i)$, the fuzzy probabilities for arrivals, in Table 6.1 and $\overline{p} = \overline{p}_1 = (0.3/0.4/0.5)$ for the fuzzy server rate. Both $\overline{p}(i)$ and $\overline{p}_1$ were used together in Chapters 7 and 9. Notice that $1 - \overline{p}_1 = (0.5/0.4/0.7)$. For the newer (faster) server we choose $\overline{p} = \overline{p}_2 = (0.5/0.6/0.7)$ so that $1 - \overline{p}_2 = (0.3/0.4/0.5)$.

As in Chapters 7 and 9 we will only compute the $\alpha = 0$ and $\alpha = 1$ cuts of the fuzzy steady state probabilities. This means that we will only have the $\alpha = 0$ and $\alpha = 1$ cuts of $\overline{U}$ and $\overline{R}$. Now both $\overline{U}$ and $\overline{R}$ are triangular shaped fuzzy numbers and we will approximate them with triangular fuzzy numbers. For example, let $\overline{R}[0] = [r_1, r_3]$ and $\overline{R}[1] = r_2$. Then we will use the triangular fuzzy number $(r_1/r_2/r_3)$ ($\approx \overline{R}$) to approximate $\overline{R}$. We will also do this for the fuzzy steady state probabilities, and $\overline{U}$, $\overline{N}$ and $\overline{X}$. We make these approximations to speed up the calculations. We will use only the alpha zero and one cuts in the rest of the book when dealing with fuzzy probabilities; we use $\alpha = 0, 1/3, 2/3, 1$ when using fuzzy arrival and service rates. In practice we recommend employing at least alpha $0, 1/3, 2/3, 1$ to get minimal approximations to $\overline{U}$, $\overline{N}$, $\overline{X}$ and $\overline{R}$. However, the calculations using $\alpha = 0, 1$ illustrate what is to be done for all α-cuts.

We will consider the eight cases shown in Table 13.1.

We will number the fuzzy numbers $\overline{U}_i$, $\overline{N}_i$, $\overline{X}_i$ and $\overline{R}_i$ according to the eight cases $i = 1, 2, 3, ..., 8$ in Table 13.1. Now we go through four steps: (1) step 1 is to find the fuzzy $\overline{p}_{ij}$, only for $\alpha = 0, 1$, in the fuzzy transition matrices $\overline{P}$; (2) calculate the fuzzy steady state probabilities, for $\alpha = 0, 1$, in Step 2 ; (3) Step 3 determines the $\overline{U}_i, ..., \overline{R}_i$, $1 \leq i \leq 8$ for alpha zero and one; and (4) Step 4 solves the minimization problem in equation (13.3) for selected values of the λ_i.

Step 1

The crisp transition matrix P was shown in Figure 6.2 for $c = 1$ and $M = 4$; and Figure 8.1 has P when $c = 2$ and $M = 4$. We discussed finding alpha-cuts of the $\bar{p}_{ij}$ in these fuzzy transition matrices in Chapters 6 through 9. What we will do here is briefly discuss finding $\alpha = 0, 1$ cuts of the $\bar{p}_{ij}$ in the fuzzy transition matrices when $c = 1$, $M = 10$ in Cases 2 and 6, and $c = 2$, $M = 10$ for Cases 4 and 8.

First let $c = 1$ and $M = 10$. The crisp transition matrix P is 11×11 and we will describe it row by row. We define the coding "a",...,"i" after the list. The rows/columns are numbered $0, 1, ..., 10$. P is

1. row #0: p(0),p(1),p(2),p(3),p(4),p(5),p(6),p(7),0,0,0;

2. row #1: a,b,c,d,e,f,g,h,i,0,0;

3. row #2: 0,a,b,c,d,e,f,g,h,i,0;

4. row #3: 0,0,a,b,c,d,e,f,g,h,i^*;

5. row #4: 0,0,0,a,b,c,d,e,f,g,h^*;

6. row #5: 0,0,0,0,a,b,c,d,e,f,g^*;

7. row #6: 0,0,0,0,0,a,b,c,d,e,f^*;

8. row #7: 0,0,0,0,0,0,a,b,c,d,e^*;

9. row #8: 0,0,0,0,0,0,0,a,b,c,d^*;

10. row #9: 0,0,0,0,0,0,0,0,a,b,c^*;

11. row #10: 0,0,0,0,0,0,0,0,0,a,b^*;

   where

1. $a = p(0)p$;

2. $b = p(1)p + p(0)(1 - p)$;

3. $c = p(2)p + p(1)(1 - p)$;

4. $d = p(3)p + p(2)(1 - p)$;

5. $e = p(4)p + p(3)(1 - p)$;

6. $f = p(5)p + p(4)(1 - p)$;

7. $g = p(6)p + p(5)(1 - p)$;

8. $h = p(7)p + p(6)(1 - p)$;

9. $i = p(7)(1 - p)$;

10. $i^* = i$;

11. $h^* = p^*(6)(1 - p) + p(7)p$;

12. $g^* = p^*(5)(1 - p) + p^*(6)p$;

13. $f^* = p^*(4)(1 - p) + p^*(5)p$;

14. $e^* = p^*(3)(1 - p) + p^*(4)p$;

15. $d^* = p^*(2)(1 - p) + p^*(3)p$;

16. $c^* = p^*(1)(1 - p) + p^*(2)p$;

17. $b^* = (1 - p) + p^*(1)p$;

where $p^*(i) = p(i) + p(i + 1) + ... + p(7)$.

We illustrated finding $\alpha = 0$ cuts of these items in Chapters 6 and 7. Many can be found by inspection, others by solving a linear programming problem and occasionally Premium Solver Platform V5.0 from Frontline Systems [3]. To show one more such calculation let us find $g^*[0]$. It is

$$g^*[0] = \{(p(5) + p(6) + p(7))(1 - p) + (p(6) + p(7))p|\mathbf{S}\}, \qquad (13.5)$$

where $\mathbf{S}$ is "$p(i) \in \overline{p}(i)[0]$, $0 \le i \le 7$, the sum of the $p(i)$ is one, $p \in \overline{p}[0]$". We see that we can choose $p(i) \in \overline{p}(i)[0]$, whose sum is one, so that $0.13 = min(p(5) + p(6) + p(7))$, $0.27 = max(p(5) + p(6) + p(7))$ and $0.06 = min(p(6) + p(7))$, $max(p(6) + p(7)) = 0.14$. If $g^*[0] = [g_1, g_2]$, then $g_1 = 0.13(1 - p) + 0.06p = 0.095$ using $p = 0.5$ if $\overline{p} = \overline{p}_1 = (0.3/0.4/0.5)$. Also, $g_2 = 0.27(1 - p) + 0.14p = 0.231$ using $p = 0.3$. Hence, $g^*[0] = [0.095, 0.231]$ in Case 2 where we use $\overline{p}_1$. Similarly, we find all the alpha zero cuts in Cases 2 and 6.

Next we look at Cases 4 and 8 where $c = 2$ and $M = 10$. We first list the rows in the 11×11 crisp transition matrix P and then define the coding "A",..., "J^*".

1. row #0: p(0),p(1),p(2),p(3),p(4),p(5),p(6),p(7),0,0,0;

2. row #1: a,b,c,d,e,f,g,h,i,0,0,;

3. row #2: A,B,C,D,E,F,G,H,I,J,0;

4. row #3: 0,A,B,C,D,E,F,G,H,I,J^*;

5. row #4: 0,0,A,B,C,D,E,F,G,H,I^*;

6. row #5: 0,0,0,A,B,C,D,E,F,G,H^*;

7. row #6: 0,0,0,0,A,B,C,D,E,F,G^*;

8. row #7: 0,0,0,0,0,A,B,C,D,E,F^*;

9. row #8: 0,0,0,0,0,0,A,B,C,D,E^*;

10. row #9: 0,0,0,0,0,0,0,A,B,C,D^*;

11. row #10: 0,0,0,0,0,0,0,0,A,B,C^*;

where

1. a,b,...,i were all defined above for $c = 1$;

2. $A = p(0)q(2|2)$;

3. $B = p(0)q(1|2) + p(1)q(2|2)$;

4. $C = p(0)q(0|2) + p(1)q(1|2) + p(2)q(2|2)$;

5. $D = p(1)q(0|2) + p(2)q(1|2) + p(3)q(2|2)$;

6. $E = p(2)q(0|2) + p(3)q(1|2) + p(4)q(2|2)$;

7. $F = p(3)q(0|2) + p(4)q(1|2) + p(5)q(2|2)$;

8. $G = p(4)q(0|2) + p(5)q(1|2) + p(6)q(2|2)$;

9. $H = p(5)q(0|2) + p(6)q(1|2) + p(7)q(2|2)$;

10. $I = p(6)q(0|2) + p(7)q(1|2)$;

11. $J = p(7)q(0|2)$;

12. $J^* = J$;

13. $I^* = p^*(6)q(0|2) + p(7)q(1|2)$;

14. $H^* = p^*(5)q(0|2) + p^*(6)q(1|2) + p(7)q(2|2)$;

15. $G^* = p^*(4)q(0|1) + p^*(5)q(1|2) + p^*(6)q(2|2)$;

16. $F^* = p^*(3)q(0|2) + p^*(4)q(1|2) + p^*(5)q(2|2)$;

17. $E^* = p^*(2)q(0|2) + p^*(3)q(1|2) + p^*(4)q(2|2)$;

18. $D^* = p^*(1)q(0|2) + p^*(2)q(1|2) + p^*(3)q(2|2)$;

19. $C^* = p^*(0)q(0|2) + p^*(1)q(1|2) + p^*(2)q(2|2)$;

$\overline{q}$	$\alpha = 0$	$\alpha = 1$
$\overline{q}(0\vert 2)[\alpha]$	$[0.25, 0.49]$	0.36
$\overline{q}(1\vert 2)[\alpha]$	$[0.42, 0.50]$	0.48
$\overline{q}(2\vert 2)[\alpha]$	$[0.09, 0.25]$	0.16

Table 13.2: $\alpha = 0, 1$ Cuts of $\overline{q}(i\vert 2)$ Using $\overline{p}_1 = (0.3/0.4/0.5)$ in Example 13.2.1.1

$\overline{q}$	$\alpha = 0$	$\alpha = 1$
$\overline{q}(0\vert 2)[\alpha]$	$[0.09, 0.25]$	0.16
$\overline{q}(1\vert 2)[\alpha]$	$[0.42, 0.50]$	0.48
$\overline{q}(2\vert 2)[\alpha]$	$[0.25, 0.49]$	0.36

Table 13.3: $\alpha = 0, 1$ Cuts of $\overline{q}(i\vert 2)$ Using $\overline{p}_2 = (0.5/0.6/0.7)$ in Example 13.2.1.1

where $p^*(i) = p(i) + p(i + 1) + ... + p(7)$.

We illustrated finding $\alpha = 0$ cuts of these items in Chapters 8 and 9. Many can be found by inspection, others by solving a linear programming problem and a few using Premium Solver Platform V5.0 from Frontline Systems [3]. To show one more. To show one more such calculation let us find $H^*[0]$. It is

$$H^*[0] = \{(p(5)+p(6)+p(7))q(0\vert 2)+(p(6)+p(7))q(1\vert 2)+p(7)q(2\vert 2)\vert \mathbf{S}\}, \quad (13.6)$$

where $\mathbf{S}$ is "$p(i) \in \overline{p}(i)[0]$, $0 \le i \le 7$, the sum of the $p(i)$ is one, $q(i\vert 2) \in \overline{q}(i\vert 2)[0]$, $1 \le i \le 3$ and the sum of the $q(i\vert 2)$ is one". We first need to compute the $\alpha = 0$ cuts of the $\overline{q}(i\vert 2)$ for all i.

The α-cuts of the $\overline{q}(i\vert 2)$ were discussed in Section 3.6.3 in Chapter 3. In Case 4 we use $\overline{p} = \overline{p}_1 = (0.3/0.4/0.5)$. Using this value of $\overline{p}$ we calculated the $\alpha = 0, 1$ cuts and they are shown in Table 13.2 . These alpha cuts would be used to get the $\alpha = 0, 1$ cuts of the $\overline{p}_{ij}$ in the fuzzy transition matrix in Case 4.

In Case 8 we use $\overline{p} = \overline{p}_2 = (0.5/0.6/0.7)$. Then the alpha cuts of the $\overline{q}(i\vert 2)$ are in Table 13.3. Now we can return to finding $H^*[0]$.

Suppose we are in Case 8. Then we use the alpha cuts of $\overline{q}(i\vert 2)$ given above for $\overline{p} = \overline{p}_2$. We may choose values of the $p(i) \in \overline{p}(i)[0]$ whose sum is one so that: (1) $min(p(5) + p(6) + p(7)) = 0.13$, $max(p(5) + p(6) + p(7)) = 0.27$; (2) $min(p(6) + p(7)) = 0.06$, $max(p(6) + p(7)) = 0.14$; and (3) $min(p(7)) = 0.03$, $max(p(7)) = 0.07$. Then we have the following linear programming problems

$$max[0.27q(0\vert 1) + 0.14q(1\vert 2) + 0.07q(2\vert 2)], \quad (13.7)$$

and

$$min[0.13q(0\vert 2) + 0.06q(1\vert 2) + 0.03q(2\vert 2)], \quad (13.8)$$

$\overline{w}$	$\alpha = 0$	$\alpha = 1$
$\overline{w}_0[\alpha]$	$[0.0000^*, 0.0000^*]$	0.0000^*
$\overline{w}_1[\alpha]$	$[0.0000^*, 0.0000^*]$	0.0000^*
$\overline{w}_2[\alpha]$	$[0.0000^*, 0.0002]$	0.0000^*
$\overline{w}_3[\alpha]$	$[0.0000^*, 0.0004]$	0.0000^*
$\overline{w}_4[\alpha]$	$[0.0000^*, 0.0013]$	0.0001
$\overline{w}_5[\alpha]$	$[0.0000^*, 0.0036]$	0.0004
$\overline{w}_6[\alpha]$	$[0.0002, 0.0103]$	0.0016
$\overline{w}_7[\alpha]$	$[0.0012, 0.0269]$	0.0069
$\overline{w}_8[\alpha]$	$[0.0110, 0.0770]$	0.0332
$\overline{w}_9[\alpha]$	$[0.0606, 0.1798]$	0.1136
$\overline{w}_{10}[\alpha]$	$[0.7134, 0.9263]$	0.8442

Table 13.4: $\alpha = 0, 1$ Cuts of the Fuzzy Steady State Probabilities, Case 4 in Example 13.2.1.1

subject to the linear constraints

$$0.09 \le q(0|2) \le 0.25, 0.42 \le q(1|2) \le 0.50, 0.25 \le q(2|2) \le 0.49, \qquad (13.9)$$

and

$$q(0|2) + q(1|2) + q(2|2) = 1. \qquad (13.10)$$

The solutions, using "simplex" in Maple [4], are: (1) $q(0|1) = 0.25, q(1|2) = 0.50, q(2|2) = 0.25$ in the *max* problem; and (2) $q(0|2) = 0.09, q(1|2) = 0.42, q(2|2) = 0.49$ in the *min* problem. Hence, we get $H^*[0] = [0.0516, 0.1550]$.

We now assume that all the alpha equal zero cuts have been found for all the elements in all the fuzzy transition matrices in Cases 1 through 8. All the fuzzy transition matrices here are fuzzy transition matrices for a fuzzy, regular, Markov chain discussed in Section 4.2 in Chapter 4.

Step 2

Here we determine (estimate) the end points of the $\alpha = 0$ cut of the fuzzy steady state probabilities for the eight cases in Table 13.1. We applied a genetic algorithm to estimate these end points and also Premium Solver Platform V5.0 from Frontline Systems [3], further details are in Chapter 19. We will not give all eight tables for the $\alpha = 0, 1$ cuts of these fuzzy steady state probabilities, so instead we present only two tables. Table 13.4 is for Case 4 and Table 13.5 has the alpha-cuts in Case 6. In these tables 0.0000* means that the number rounded off to four decimal places is 0.0000.

The $\overline{w}_i[1]$, $0 \le i \le M$, are easy to get since the $\alpha = 1$ cut of the fuzzy transition matrix is just a crisp matrix P, whose $p_{ij} \in [0, 1]$ and the row sums

$\overline{w}$	$\alpha = 0$	$\alpha = 1$
$\overline{w}_0[\alpha]$	$[0.0000^*, 0.0000^*]$	0.0000^*
$\overline{w}_1[\alpha]$	$[0.0000^*, 0.0000^*]$	0.0000^*
$\overline{w}_2[\alpha]$	$[0.0000^*, 0.0000^*]$	0.0000^*
$\overline{w}_3[\alpha]$	$[0.0000^*, 0.0000^*]$	0.0000^*
$\overline{w}_4[\alpha]$	$[0.0000^*, 0.0000^*]$	0.0000^*
$\overline{w}_5[\alpha]$	$[0.0000^*, 0.0000^*]$	0.0000^*
$\overline{w}_6[\alpha]$	$[0.0000^*, 0.0003]$	0.0000^*
$\overline{w}_7[\alpha]$	$[0.0001, 0.0020]$	0.0005
$\overline{w}_8[\alpha]$	$[0.0017, 0.0153]$	0.0057
$\overline{w}_9[\alpha]$	$[0.0406, 0.1155]$	0.0727
$\overline{w}_{10}[\alpha]$	$[0.8676, 0.9576]$	0.9211

Table 13.5: $\alpha = 0, 1$ Cuts of the Fuzzy Steady State Probabilities, Case 6 in Example 13.2.1.1

α	$\overline{U}_4[\alpha]$	$\overline{R}_4[\alpha]$
1	1.0000	12.2379
0	$[1.0000, 1.0000]$	$[9.5270, 16.5205]$

Table 13.6: Alpha Cuts of $\overline{U}$ and $\overline{R}$, Case 4 in Example 13.2.1.1

of P are all one. Therefore P^n converges as $n \to \infty$. The $\overline{w}_i[1]$, $0 \le i \le M$, are in P^n for sufficiently large n.

Step 3

Now we get the $\alpha = 0, 1$ cuts of $\overline{U}$, $\overline{N}$, $\overline{X}$ and $\overline{R}$. When $c = 1$ the necessary computations were outlined in Step 3 in Chapters 6 and 7. Equations (7.4) and (7.5) of Chapter 7 were for $\overline{p} = \overline{p}_1 = (0.3/0.4/0.5)$. We need to change the "p-values" in these equations when employing $\overline{p} = \overline{p}_2 = (0.5/0.6/0.7)$.

For $c = 2$ the needed calculations were discussed in Step 3 in Chapters 8 and 9. The "p-values" in equations (9.4)-(9.6) of Chapter 9 need to be chosen from $\overline{p}_1$ or $\overline{p}_2$.

For our optimization models we will need only the $\alpha = 0, 1$ cuts of $\overline{U}$ and $\overline{R}$ for all the eight cases in Table 13.1. Two of these results are shown in the following two tables, Table 13.6 and Table 13.7.

α	$\overline{U}_7[\alpha]$	$\overline{R}_7[\alpha]$
1	0.9700	3.0363
0	[0.9089, 0.9924]	[2.2951, 4.0154]

Table 13.7: Alpha Cuts of $\overline{U}$ and $\overline{R}$, Case 7 in Example 13.2.1.1

Case	L_{z1}	m_{z1}	R_{z1}
1	1.0308	9.8690	1.6908
2	2.5309	24.8690	4.1908
3	0.6024	4.7800	0.9282
4	1.3554	12.2379	2.1413
5	0.5140	6.5240	0.6580
6	1.2294	16.5240	1.6937
7	0.3706	3.0362	0.4896
8	0.7520	7.9389	0.9260

Table 13.8: Central Value and Certain Areas Under the Graph of $\overline{R}_i$, the Eight Cases in Example 13.2.1.1

Step 4

The first thing to do is compute L_z and R_z, for $\overline{R}_i$, $i = 1, 2, ..., 8$. Recall that L_z (R_z) is the area under the graph of the membership function for $\overline{R}$ to the left (right) of m_z. These are needed for the objective function to be minimized. But these calculations are easy since we are approximating $\overline{R}$ with a triangular fuzzy number. Suppose $(a/b/c)$ is the triangular fuzzy number used to approximate $\overline{R}$. Then $L_z = 0.5(b - c)$ and $R_z = 0.5(c - b)$ because these regions are right triangles of height one. The results are shown in Table 13.8. The columns in Table 13.8 are labeled L_{z1}, m_{z1}, R_{z1} to compare to those for $\overline{U}$ computed in the next section, which will be labeled L_{z2}, m_{z2}, R_{z2}, and placed into Table 13.10.

Finally, we get to the objective function

$$\lambda_1(3 - L_{z1i}) + \lambda_2 m_{z1i} + \lambda_3 R_{z1i}, \tag{13.11}$$

to be minimized. This is equation (13.3) with $K_1 = 3$, we may use this value of K_1 since all the values of L_{z1} in Table 13.8 are between zero and 2.5309. Given values of the λ_i, substitute the L_{z1i}, m_{z1i} and R_{z1i} from Table 13.8, we find the value of i (Case number in Table 13.1) that makes the expression in equation (13.11) a minimum. The results for three choices for the λ values is shown in Table 13.9. The (a, b, c) labels at the top of this table means $\lambda_1 = a$, $\lambda_2 = b$ and $\lambda_3 = c$.

We see from Table 13.9 that the optimal solution in all cases is Case

Case	$(1/3, 1/3, 1/3)$	$(0.4, 0.4, 0.2)$	$(0.3, 0.5, 0.2)$
1	4.5097	5.0734	5.8634
2	9.8430	10.9734	13.4134
3	2.7020	3.0567	3.2949
4	5.3412	5.9812	7.0406
5	3.2227	3.7356	4.1394
6	6.6628	7.6566	9.1319
7	2.0518	2.3642	2.4049
8	3.7043	4.2600	4.8290

Table 13.9: The Results of $min\overline{R}$ for the Eight Cases in Example 13.2.1.1

7 ($c = 2$, $M = 4$, $\overline{p} = (0.5/0.6/0.7)$) from Table 13.1. In Case 7 we found that $\overline{N}[0] = [3.1994, 3.7454]$ so the system is usually full of customers and we probably have too many lost customers. We will add the objective of minimizing lost customers in Chapter 17.

13.2.2 Minimize $\overline{R}$ and Maximize $\overline{U}$

Let $\overline{Z1} = \overline{R}(M, c, \overline{p})$ and $\overline{Z2} = \overline{U}(M, c, \overline{p})$. Let the computations discussed above produce L_{zi}, m_{zi} and R_{zi} where we use $i = 1$ for $\overline{Z1}$ and $i = 2$ for $\overline{Z2}$.

We wish to maximize $\overline{U}$ which means minimize L_{z2}, maximize m_{z2} and maximize R_{z2}. We will turn these maximization goals around to be minimization objectives to be more compatible with minimize $\overline{R}$. Let K_2 (K_3) be a positive constant so that max m_{z2} (max R_{z2}) is equivalent to min $(K_2 - m_{z2})$ (min $(K_3 - R_{z2})$). Then for $\tau_i > 0$, $\tau_1 + \tau_2 + \tau_3 = 1$, in place of $max\overline{U}$ we use minimize

$$V2 = \tau_1 L_{z2} + \tau_2(K_2 - m_{z2}) + \tau_3(K_3 - R_{z2}). \tag{13.12}$$

Also let, from equation (13.3), us minimize

$$V1 = \lambda_1(K_1 - L_{z1}) + \lambda_2 m_{z1} + \lambda_3 R_{z1}. \tag{13.13}$$

First the decision maker chooses μ_1 and μ_2 for the two primary goals of $minV1$ and $minV2$ with $\mu_i > 0$, $\mu_1 + \mu_2 = 1$. We start with

$$min[\mu_1(V1) + \mu_2(V2)]. \tag{13.14}$$

Then we choose $\lambda_i > 0$, $\lambda_1 + \lambda_2 + \lambda_3 = 1$ and $\tau_i > 0$, $\tau_1 + \tau_2 + \tau_3 = 1$, and substitute $V1$ and $V2$ into equation (13.14) to obtain our overall objective function to minimize. The constraints on the variables are those in equation (13.4). Of course, we would vary the μ_i, λ_i and the τ_i to obtain a collection of solutions to discuss with management.

We would really need to compute $\overline{w}_i[\alpha]$, $0 \leq i \leq M$, for at least $\alpha = 0, 1/3, 2/3, 1.0$ in order to get a minimal approximation to the fuzzy numbers $\overline{R}$ and $\overline{U}$ so that we can get an approximations, using the Trapezoidal Rule (see Example 13.3.1.1 below), to the areas L_{zi} and R_{zi}. However, to simplify the discussion and to cut back on the amount of calculations, we are using only $\alpha = 0, 1$ in our model.

In this section we only looked at two possible goals: $max[\overline{U} = $ server utilization]; and $min[\overline{R} = $ average response time]. However, other goals can be investigated. For example, we might consider minimizing the number of lost customers (turned away because of finite system capacity) per unit time. We first find the crisp expression for the expected number of requests (customers) rejected per unit time due to finite system capacity and then fuzzify it. Now $w_M \times 100$, w_M is the probability the system is full, is the percent of requests rejected per unit time and let Λ be the expected number of customers arriving at the system per unit time. Then the expected number of customers lost per unit time (LC) would be $w_M \Lambda$. It follows that

$$LC = w_M \Lambda = w_M \sum_{i=0}^{L} ip(i), \qquad (13.15)$$

where $p(i)$ is the probability that i customers arrive during the unit time interval and we have assumed that $p(i) = 0$ for $i > L$. Hence

$$\overline{LC} = [\overline{w}_M][\sum_{i=0}^{L} i\overline{p}(i)], \qquad (13.16)$$

to be evaluated by α-cuts and restricted fuzzy arithmetic. We get $\overline{w}_M$ from the fuzzy steady state probabilities and the $\overline{p}(i)$ are calculated from data producing a table as in Table 6.1. M is system capacity and $\overline{w}_M$ is the fuzzy steady state probability that the system (servers and queue) is full. Then we have a new goal $min\overline{LC}$. Alternatively, we could consider $min\overline{w}_M$. We will add the goal of minimizing $\overline{LC}$ in Chapter 17.

Example 13.2.2.1

We consider the same eight cases in Table 13.1 with the same two values of $\overline{p}$. The objective is given in equation (13.14) with $V1$ from equation (13.13) and $V2$ in equation (13.12). In Example 13.2.1.1 we calculated L_{z1} and R_{z1} for $\overline{R}$ in the eight cases, so now we need to do the same for $\overline{U}$. We found the alpha-cuts of $\overline{U}$ for all the cases in Example 13.2.1.1 and it is easy to compute L_{z2} and R_{z2} since these are just areas of right triangles whose height is one. The results are shown in Table 13.10. Let us explain how we obtained $L_{z2} = R_{z2} = 0$ and $m_{z2} = 1$ in Cases 1,2,4 and 6. In all these cases we got $\overline{U}[0] = [1, 1]$ and $\overline{U}[1] = 1$. For example in Case 6 where $c = 1$ and

Case	L_{z2}	m_{z2}	R_{z2}
1	0	1.0000	0
2	0	1.0000	0
3	0.0112	0.9927	0.0030
4	0	1.0000	0
5	0.0001	1.0000	0
6	0	1.0000	0
7	0.0306	0.9700	0.0112
8	0.0003	0.9999	0.0000*

Table 13.10: Central Value and Certain Areas Under the Graph of $\overline{U}_i$, the Eight Cases in Example 13.2.2.1

Case	I	II	III	IV	V	VI
1	2.9725	3.3641	3.8380	2.2039	2.5094	2.8254
2	6.1724	6.9040	8.3680	4.3372	4.8694	5.8454
3	1.8899	2.1561	2.2990	1.4839	1.7058	1.8010
4	3.4714	3.9087	4.5443	2.5365	2.8725	3.2962
5	2.2003	2.5614	2.8036	1.6891	1.9743	2.1358
6	4.2643	4.9139	5.7991	3.0651	3.5426	4.1328
7	1.5043	1.7458	1.7700	1.2306	1.4365	1.4526
8	2.4893	2.8760	3.2175	1.8818	2.1840	2.4117

Table 13.11: Final Results of $min\overline{R}$ and $max\overline{U}$ in Example 13.2.2.1

$M = 10$ we obtained $\overline{w}_0[0] = [0,0]$ and $\overline{w}_0[1] = 0$, there this zero was 0.0000 rounded off to four decimal places, which implies that $w_1 + ... + w_{10} = 1$ for $w_i \in \overline{w}_i[0]$, $1 \leq i \leq 10$. From this last result we see that all the alpha-cuts of $\overline{U}$ equal the real number one.

Once we have these areas, it is easy to calculate the value of the objective function for various values of the parameters. Selections of the μ_i, λ_i and τ_i giving six cases are shown in Table 13.11. The cases I,II,...,VI across the top of the table are: (1) I is $\mu_1 = 0.6, \mu_2 = 0.4, \lambda_i = \tau_i = 1/3$ all i; (2) II equals $\mu_1 = 0.6, \mu_2 = 0.4, \lambda_1 = \lambda_2 = \tau_2 = \tau_3 = 0.4, \lambda_3 = \tau_1 = 0.2$; (3)case III has $\mu_1 = 0.6, \mu_2 = 0.4, \lambda_1 = \tau_3 = 0.3, \lambda_2 = \tau_2 = 0.5, \lambda_3 = \tau_1 = 0.2$; (4) IV is case I with $\mu_1 = 0.4, \mu_2 = 0.6$; (5) V is case II with $\mu = 0.4, \mu_2 = 0.6$; and (6) VI has the same as case III except $\mu_1 = 0.4, \mu_2 = 0.6$. In equation (13.12) for $V2$ we used $K_2 = 2$ since the maximum value of $m_{z2} = 1$ and $K_3 = 1$ because $R_{z2} \leq 0.0030$ from Table 13.10.

Being a minimization problem we see from Table 13.11 that in all cases the solution is Case 7 as in Example 13.2.1.1.

$\overline{R}$	Triangular Fuzzy Number
$\overline{R}_1$	(7.8/9.9/13.2)
$\overline{R}_2$	(19.8/24.9/33.2)
$\overline{R}_3$	(3.6/4.8/6.6)
$\overline{R}_4$	(9.5/12.2/16.5)
$\overline{R}_5$	(5.5/6.5/7.8)
$\overline{R}_6$	(14.1/16.5/19.9)
$\overline{R}_7$	(2.3/3.0/4.0)
$\overline{R}_8$	(6.4/7.9/9.8)

Table 13.12: The Triangular Fuzzy Numbers for the $\overline{R}$ in Example 13.2.3.1

13.2.3 Ranking the Fuzzy Sets

We discussed ranking fuzzy numbers in Section 2.6 of Chapter 2. Let us apply this to $min\overline{R}$ and to $max\overline{U}$. We first consider only $min\overline{R}$. Recall that the highest ranked fuzzy numbers are in the set H_K and the lowest ranked fuzzy numbers are in H_1.

Fuzzy sets $\overline{A}_i$ are the values of $\overline{R}$ for $M = 4, 5, ..., 10$, $c = 1, 2, 3$ and $\overline{p} = \overline{p}_i$, $i = 1, 2$. Since we wish to minimize $\overline{R}$ we look at all the fuzzy sets in H_1. A problem may arise when we get more than one $\overline{R}$ in H_1. In general we suggest presenting management with all solutions in H_1 and let them pick one to implement. If we require a unique solution, then we could go back to the previous section to resolve the conflict.

Example 13.2.3.1

We first round off the $\alpha = 0$ cut and the $\alpha = 1$ cut of $\overline{R}_i$, $1 \leq i \leq 8$, to one decimal place for ease in producing a graph of these fuzzy numbers. Since we only have the $\alpha = 0$ and $\alpha = 1$ cut for these fuzzy numbers we will approximate them by triangular fuzzy numbers. For example, if the alpha equal zero cut of a $\overline{R}$ is $[r_1, r_3]$ and the alpha equal one cut is r_2, then $\overline{R} \approx (r_1/r_2/r_3)$. We will now drop the $\approx$ and write a $\overline{R}$ as a triangular fuzzy number. These triangular fuzzy numbers for all the $\overline{R}$ are shown in Table 13.12.

Next we graph these triangular fuzzy number from Table 13.12 to see how they line up from smallest to largest. This is now shown in Figure 13.1. We want H_1 the set of smallest fuzzy numbers. From Figure 13.1 clearly H_1 consists of $\overline{R}_7$ and possibly also $\overline{R}_3$. From Section 2.6 of Chapter 2, and Figure 2.6, we get $v(\overline{R}_7, \overline{R}_3) \approx 0.2 < 0.8$ so we conclude $\overline{R}_7 < \overline{R}_3$ and H_1 has only $\overline{R}_7$. For the comparison of fuzzy numbers we will use $\eta = 0.8$ as discussed in Section 2.6 of Chapter 2. If we draw a horizontal line at 0.8 on the vertical axis we that the height of the intersection of $\overline{R}_7$ and $\overline{R}_3$ is below

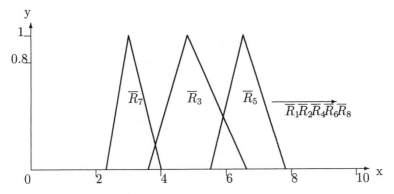

Figure 13.1: Ranking the Fuzzy Numbers $\overline{R}$ in Example 13.2.3.1

this line. Hence $\overline{R}_7 < \overline{R}_3$. So from ranking the fuzzy numbers we get Case 7 as the optimal solution, the same as in Example 13.2.1.1.

Next consider maximizing $\overline{U}$ and minimizing $\overline{R}$. We obtain fuzzy sets $\overline{A}_{i1}$ for $\overline{R}$ and $\overline{A}_{i2}$ for $\overline{U}$. Find H_K for $\overline{U}$ and H_1 for $\overline{R}$. Hopefully there is a $\overline{A}_{i1} \in H_1$ for $M = M^*$, $c = c^*$ and $\overline{p} = \overline{p}^*$ and a $\overline{A}_{i2} \in H_K$ for the same values of the variables. Then these values of the variables is optimal. There may be other values of the variables that are optimal, then we may have to resolve the conflict for a unique solution. However, there may be no optimal values of the variables and then we abandon this procedure and go to Section 13.2.

Example 13.2.3.2

We first round off the $\alpha = 0$ cut and the $\alpha = 1$ cut of $\overline{U}_i$, $1 \le i \le 8$, to two decimal places for ease in producing a graph of these fuzzy numbers. We now use two decimal places because otherwise, using only one decimal place, all results (except one) would equal 1.0. Since we only have the $\alpha = 0$ and $\alpha = 1$ cut for these fuzzy numbers we will approximate them by triangular fuzzy numbers. For example, if the alpha equal zero cut of a $\overline{U}$ is $[u_1, u_3]$ and the alpha equal one cut is u_2, then $\overline{U} \approx (u_1/u_2/u_3)$. We will now drop the $\approx$ and write a $\overline{U}$ as a triangular fuzzy number. These triangular fuzzy numbers for all the $\overline{U}$ are shown in Table 13.13.

In Table 13.13 $(1.00/1.00/1.00)$ represents the (crisp) real number one. The graph of the triangular fuzzy numbers in Table 13.13 are shown in Figure 13.2. The large "dot" in this figure at position $(1.00, 1)$ represents the "triangular" fuzzy number (real number) $(1.00/1.00/1.00)$ for $\overline{U}_1 = \overline{U}_2 = \overline{U}_4 = \overline{U}_5 = \overline{U}_6 = \overline{U}_8$. It is easy to determine that $\overline{U}_7 < \overline{U}_3 < \overline{U}_i$ for $i = 1, 2, 4, 5, 6, 8$ using $\eta = 0.8$ from Section 2.6 in Chapter 2. In fact the

$\overline{U}$	Triangular Fuzzy Number
$\overline{U}_1$	(1.00/1.00/1.00)
$\overline{U}_2$	(1.00/1.00/1.00)
$\overline{U}_3$	(0.97/0.99/1.00)
$\overline{U}_4$	(1.00/1.00/1.00)
$\overline{U}_5$	(1.00/1.00/1.00)
$\overline{U}_6$	(1.00/1.00/1.00)
$\overline{U}_7$	(0.91/0.97/0.99)
$\overline{U}_8$	(1.00/1.00/1.00)

Table 13.13: The Triangular Fuzzy Numbers for the $\overline{U}$ in Example 13.2.3.1

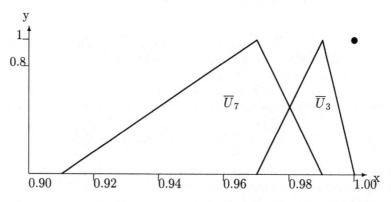

Figure 13.2: Ranking the Fuzzy Numbers $\overline{U}$ in Example 13.2.3.2

height of the intersection between $\overline{U}_7$ and $\overline{U}_3$ is 0.5, see Figure 2.6. Also the height of the intersection between $\overline{U}_3$ and the $\overline{U}_i$, $i = 1, 2, 4, 5, 6, 8$ is zero. In this case we want H_3 the set of largest fuzzy numbers. The set of $\overline{U}_i$, $1 \leq i \leq 8$ breaks up into three sets: $H_1 = \{\overline{U}_7\}$ the smallest set; $H_2 = \{\overline{U}_3\}$; and the largest set $H_3 = \{\overline{U}_i | i = 1, 2, 4, 5, 6, 8\}$. Notice that the height of the intersection of $\overline{U}_7$ and $\overline{U}_3$ lies below the horizontal line at 0.8 on the y−axis. Hence $\overline{U}_7 < \overline{U}_3$. The optimal solution is Cases 1,2,4,5,6 and 8.

Comparing the results on $\overline{R}$ and $\overline{U}$ we have a conflict. $\overline{R}_7$ was the smallest of the $\overline{R}_i$ but $\overline{U}_7$ does not belong to the largest set of the $\overline{U}_i$. In fact, $\overline{U}_7$ turns out to be the smallest of the $\overline{U}_i$. To resolve this conflict we may turn to the analytical method in Example 13.2.2.1 were the optimal solution was Case 7.

Too often we will end up with a conflict, using the method of ranking fuzzy numbers, when we have two, or more, goals. Here we had two goals: $min\overline{R}$ and $max\overline{U}$. In Chapter 17 we will have three goals. If we have conflicting goals, them we should end up with conflicting solutions when we apply the ranking of fuzzy sets method to both goals. So, when we have two, or more, conflicting goals we will not employ the method of ranking fuzzy numbers in the rest of this book. However, it is still a good idea to graph all the fuzzy numbers, for each goal, to see how they line up from smallest to largest, as long as the total number of fuzzy numbers for each goal is not too great.

13.3 Fuzzy Arrival/Service Rates

As in the previous section we want to $max\overline{U}$ and $min\overline{R}$. The variables are $c = 1, 2, 3$, $M = 4, 5, ..., 10$ and two possible values of $\overline{\mu}$. Assume that $\overline{\lambda}$, the fuzzy arrival rate, has been estimated and it is fixed. The two values of $\overline{\mu}$ are $\overline{\mu}_1$ and $\overline{\mu}_2$. The $\overline{\mu}_1$ is for the servers we are now using and the new $\overline{\mu}_2$ is for faster servers we are considering using. Both values of $\overline{\mu}$ will be triangular fuzzy numbers, but $\overline{\mu}_2$ is $\overline{\mu}_1$ shifted to the right, for a faster server. We obtained $\overline{\mu}_1$ from data, as shown in Chapter 3, but having no experience with the new server $\overline{\mu}_2$ was determined from "expert" opinion. All servers are the same which means they all have property $\overline{\mu}_1$ or they all have $\overline{\mu}_2$.

We will now present two methods of optimization. The first one is to minimize the fuzzy set $\overline{R}$, or maximize $\overline{U}$, and the second one ranks the fuzzy sets from smallest to largest.

13.3.1 Minimize $\overline{R}$

We will first consider minimizing $\overline{R}$, a similar procedure can be employed for maximizing $\overline{U}$, and then we will discuss optimizing both.

Now $\overline{R}$ is a function of c, M and $\overline{\mu}$ and we show this as writing $\overline{Z} = \overline{R}(M, c, \overline{\mu})$. So, for different values of the variables we get triangular shaped fuzzy numbers $\overline{Z}$. Find the values of the variables to minimize $\overline{Z}$

We discussed the problem of minimizing a fuzzy set in Section 2.5 of Chapter 2, so we will only give a brief discussion here. For our problem let: (1) m_z be the center of the core of $\overline{Z}$ (the core of a fuzzy number is the interval where the membership function equals one); (2) L_z be the area under the graph of the membership function to the left of m_z; and (3) R_z be the area under the graph of the membership function to the right of m_z. See Figure 2.5. In our application the core of $\overline{Z}$ will be a single point. For $min\overline{R}$ we substitute: (1) $min[m_z]$; (2) $maxL_z$, or maximize the possibility of obtaining values less than m_z; and (3) $minR_z$, or minimize the possibility of obtaining values greater than m_z. So for $min\overline{R}$ we have

$$V = (maxL_z, min[m_z], minR_z). \tag{13.17}$$

First let K_1 be a sufficiently large positive number so that $maxL_z$ is equivalent to $minL_z^*$ where $L_z^* = K_1 - L_z$. The multiobjective problem is

$$minV = (minL_z^*, min[m_z], minR_z). \tag{13.18}$$

One way to explore the undominated set (see Section 2.5) is to change the multiobjective problem into a single objective. The single objective problem is

$$min(\lambda_1[K_1 - L_z] + \lambda_2 m_z + \lambda_3 R_z), \qquad (13.19)$$

where $\lambda_i > 0$, $1 \le i \le 3$, $\lambda_1 + \lambda_2 + \lambda_3 = 1$, and the values of the variables are

$$c = 1, 2, 3; \quad M = 4, 5, ..., 10; \quad \overline{\mu} = \overline{\mu}_i, i = 1, 2. \qquad (13.20)$$

You will get different undominated solutions by choosing different values of $\lambda_i > 0$, $\lambda_1 + \lambda_2 + \lambda_3 = 1$. The decision maker is to choose the values of the weights λ_i for the three minimization goals. Usually one picks different values for the λ_i to explore the solution set and then lets the decision maker choose an optimal solution from this set of solutions. It is usually best to present management with a number of optimal solutions, instead of only one optimal solution. Managers are decision makers, and when they have multiple solutions to pick from, they can decide on one of them weighing the various alternatives associated with each.

This is how we propose to handle the problem of $min\overline{R}$ in finding the optimal system. Numerical solutions to this optimization problem for continuous variables can be difficult. In the past we have employed an evolutionary algorithm to generate good approximate solutions in the continuous case. However, here we are in a discrete situation ($c = 1, 2, M = 4, 10$,etc.).

Example 13.3.1.1

Let us set $\overline{\lambda} = (3/4/5)$, for the fuzzy arrival rate, which was used in Chapter 12 and, for the normal fuzzy server rate, have $\overline{\mu} = \overline{\mu}_1 = (5/6/7)$, also used in Chapter 12. The faster fuzzy server rate will be $\overline{\mu}_2 = (6/7/8)$. The speed of the server is measured with respect to the time interval δ.

We will consider the eight cases in Table 13.14. These are similar to the eight cases used in Example 13.2.1.1. We will number the fuzzy numbers $\overline{U}_i$, $\overline{N}_i$, $\overline{X}_i$ and $\overline{R}_i$ according to the eight cases $i = 1, 2, 3, ..., 8$ in Table 13.14. Now we go through three steps: (1) find the fuzzy steady state probabilities in Step 1; (2) Step 2 determines the $\overline{U}_i, ..., \overline{R}_i$, $1 \le i \le 8$; and (3) Step 3 solves the minimization problem in equations (13.19) and (13.20) for selected values of the λ_i.

Step 1

We calculated the alpha-cuts $\overline{w}_i[\alpha]$ of the fuzzy steady state probabilities only for $\alpha = 0, 1/3, 2/3, 1$ to obtain a minimal approximation to the fuzzy numbers $\overline{U}_i, ..., \overline{R}_i$, $1 \le i \le 8$. These calculations, for cases 1,3,5 and 7, were discussed in Chapter 12. So let us discuss in detail only the needed computations for cases 2,4,6 and 8.

Case	Number of Servers	System Capacity	Server Rate
1	$c = 1$	$M = 4$	$\overline{\mu}_1$
2	$c = 1$	$M = 10$	$\overline{\mu}_1$
3	$c = 2$	$M = 4$	$\overline{\mu}_1$
4	$c = 2$	$M = 10$	$\overline{\mu}_1$
5	$c = 1$	$M = 4$	$\overline{\mu}_2$
6	$c = 1$	$M = 10$	$\overline{\mu}_2$
7	$c = 2$	$M = 4$	$\overline{\mu}_2$
8	$c = 2$	$M = 10$	$\overline{\mu}_2$

Table 13.14: The Eight Cases in Example 13.3.1.1

Let $c = 1$ and $M = 10$. The crisp probabilities, which are similar to equations (12.1) and (12.2), are [5]

$$w_n = \rho^n w_0, \tag{13.21}$$

for $1 \leq n \leq 10$, where $\rho = \lambda/\mu$, and

$$w_0 = \begin{cases} \frac{1-\rho}{1-\rho^{11}}, & \rho \neq 1 \\ \frac{1}{11}, & \rho = 1. \end{cases} \tag{13.22}$$

We calculate the fuzzy steady state probabilities by their α-cuts. So

$$\overline{w}_n[\alpha] = \{\rho^n \frac{1-\rho}{1-\rho^{11}} | \ \mathbf{S} \ \}, \tag{13.23}$$

for $n = 0, 1, 2, 3, ..., 10$, assuming $\rho \neq 1$, and the statement $\mathbf{S}$ is "$\lambda \in \overline{\lambda}[\alpha]$, $\mu \in \overline{\mu}[\alpha]$", for all α in $[0,1]$. We determined $\overline{w}_i[\alpha]$, $0 \leq i \leq 10$ and $\alpha = 0, 1/3, 2/3$ using the non-linear optimizer discussed after Example 12.2.1 in Chapter 12. This covers cases 2 and 6.

Next consider $c = 2$ and $M = 10$. The crisp probabilities, which are like equations (12.13) and (12.14), are [5]

$$w_0 = [1 + \rho + \frac{\rho^2(1 - \{\rho/2\}^9)}{2(1 - \rho/2)}]^{-1}, \tag{13.24}$$

if $\rho \neq 2$ and

$$w_0 = [1 + \rho + 4.5\rho^2]^{-1}, \tag{13.25}$$

when $\rho = 2$. Then

$$w_n = \begin{cases} \frac{\rho^n}{n!} w_0, & 0 \leq n \leq 2 \\ \\ \frac{\rho^n}{2^{n-1}} w_0, & 2 \leq n \leq 10. \end{cases} \tag{13.26}$$

$\overline{w}$	Alpha Zero Cut
$\overline{w}_0[0]$	$[0.0909, 0.5715]$
$\overline{w}_1[0]$	$[0.0909, 0.2501]$
$\overline{w}_2[0]$	$[0.0909, 0.1501]$
$\overline{w}_3[0]$	$[0.0450, 0.1120]$
$\overline{w}_4[0]$	$[0.0193, 0.0956]$
$\overline{w}_5[0]$	$[0.0083, 0.0909]$
$\overline{w}_6[0]$	$[0.0035, 0.0909]$
$\overline{w}_7[0]$	$[0.0015, 0.0909]$
$\overline{w}_8[0]$	$[0.0006, 0.0909]$
$\overline{w}_9[0]$	$[0.0003, 0.0909]$
$\overline{w}_{10}[0]$	$[0.0001, 0.0909]$

Table 13.15: Alpha Zero Cuts of the Fuzzy Steady State Probabilities, Case 2 in Example 13.3.1.1

Then we find $\overline{w}_i[\alpha]$, $0 \leq i \leq 10$, for $\alpha = 0, 1/3, 2/3$ as discussed above using the non-linear optimizer. This completes cases 4 and 8.

We will not present tables for all these alpha-cuts for all of these fuzzy steady state probabilities. Instead we give only two tables, Table 13.15 for Case 2 and Table 13.16 for Case 8, for $\alpha = 0$. In Table 13.16 0.0000* means that it is 0.0000 rounded to four decimal places.

Step 2

Here we are to get the alpha-cuts, $\alpha = 0, 1/3, 2/3$, of the $\overline{U}_i$, $\overline{N}_i$, $\overline{X}_i$ and $\overline{R}_i$, $1 \leq i \leq 8$. We do not need the $\overline{U}_i$ in this example but it will be used in the following example, Example 13.3.2.1. In all cases we find the alpha-cuts of $\overline{U}_i$, $\overline{N}_i$ and $\overline{X}_i$ by solving a linear programming problem as discussed in Chapter 12. We just change the constraints, equation (12.8) or (12.21), to be

$$w_{i1}(\alpha) \leq w_i \leq w_{i2}(\alpha), 0 \leq i \leq M, w_0 + ... + w_M = 1, \tag{13.27}$$

for $\alpha = 0, 1/3, 2/3$, $M = 4$ or $M = 10$. where $\overline{w}_i[\alpha] = [w_{i1}(\alpha), w_{i2}(\alpha)]$ all i and all α.

There will be another change in computing the $\overline{X}_i$. Let $\overline{\mu}_i[\alpha] = [\mu_{i1}(\alpha), \mu_{i2}(\alpha)]$ for $i = 1, 2$ and $\alpha = 0, 1/3, 2/3$. If $c = 1$, then: (1) in equation (12.11) use $\mu_{i1}(\alpha)$ for the "5"; and (2) in equation (12.12) use $\mu_{i2}(\alpha)$ for the "7". If $c = 2$, then: (1) in equation (12.24) use $\mu_{i1}(\alpha)$ for the "5" and two times this μ value for the "10"; and (2) in equation (12.25) use $\mu_{i2}(\alpha)$ for the "7" and twice this μ value for the "14". Of course, we need to do this when alpha is $0, 1/3, 2/3$ and $i = 1, 2$. For example, consider case 7 and $\alpha = 2/3$. Then the linear programming problem to solve for the left end

$\overline{w}$	Alpha Zero Cut
$\overline{w}_0[0]$	$[0.4118, 0.6842]$
$\overline{w}_1[0]$	$[0.2566, 0.3432]$
$\overline{w}_2[0]$	$[0.0481, 0.1430]$
$\overline{w}_3[0]$	$[0.0090, 0.0596]$
$\overline{w}_4[0]$	$[0.0017, 0.0248]$
$\overline{w}_5[0]$	$[0.0003, 0.0103]$
$\overline{w}_6[0]$	$[0.0001, 0.0043]$
$\overline{w}_7[0]$	$[0.0000^*, 0.0018]$
$\overline{w}_8[0]$	$[0.0000^*, 0.0007]$
$\overline{w}_9[0]$	$[0.0000^*, 0.0003]$
$\overline{w}_{10}[0]$	$[0.0000^*, 0.0001]$

Table 13.16: Alpha Zero Cuts of the Fuzzy Steady State Probabilities, Case 8 in Example 13.3.1.1

α	$\overline{U}_4[\alpha]$	$\overline{R}_4[\alpha]$
1	0.1615	0.1912
2/3	$[0.1371, 0.2100]$	$[0.1304, 0.2732]$
1/3	$[0.1190, 0.2646]$	$[0.0915, 0.4051]$
0	$[0.1157, 0.3233]$	$[0.0641, 0.6189]$

Table 13.17: Alpha Cuts of $\overline{U}$ and $\overline{R}$, Case 4 in Example 13.3.1.1

point of $\overline{X}_7[2/3]$ has objective function

$$6.6667(w_1) + 2(6.6667)(w_2 + ... + w_4), \tag{13.28}$$

to be minimized subject to the constraints given above. For the right end point the objective function is

$$7.3333(w_1) + 2(7.3333)(w_2 + ... + w_4), \tag{13.29}$$

to be maximized subject to the same constraints. Table 13.17 shows the results for $\overline{U}_i$ and $\overline{R}_i$ when $i = 4$ and Table 13.18 gives the alpha-cuts if $i = 7$.

Step 3

The first thing to do is compute L_z and R_z, defined above, for $\overline{R}_i$, $i = 1, 2, ..., 8$. These are needed for the objective function to be minimized. From the α-cuts we have seven points on the graph of the membership function for $\overline{R}_i$. For a given value of i, let those seven points be

α	$\overline{U}_7[\alpha]$	$\overline{R}_7[\alpha]$
1	0.1244	0.1531
2/3	[0.0986, 0.1548]	[0.1116, 0.2112]
1/3	[0.0769, 0.1903]	[0.0815, 0.2932]
0	[0.0588, 0.2315]	[0.0595, 0.4112]

Table 13.18: Alpha Cuts of $\overline{U}$ and $\overline{R}$, Case 7 in Example 13.3.1.1

Case	L_{z1}	m_{z1}	R_{z1}
1	0.1240	0.3358	0.3019
2	0.2188	0.4705	0.7303
3	0.0666	0.1824	0.1445
4	0.0747	0.1912	0.1699
5	0.0964	0.2652	0.2007
6	0.1427	0.3288	0.4367
7	0.0504	0.1531	0.1114
8	0.0533	0.1554	0.1091

Table 13.19: Central Value and Certain Areas Under the Graph of $\overline{R}_i$, the Eight Cases in Example 13.3.1.1

$(x_1, 0), (x_2, 1/3), (x_3, 2/3), (x_4, 1), (x_5, 2/3), (x_6, 1/3), (x_7, 0)$. For example, $[x_1, x_7]$ is the $\alpha = 0$ cut, $[x_2, x_6]$ is the $\alpha = 1/3$ cut, etc. To approximate the graph of the membership function, we connect adjacent pairs of points by straight line segments. Then we find the area under this graph and over the interval $[x_1, x_4]$ to approximate L_z. The area under the graph, using straight line segments, and over $[x_4, x_7]$ approximates R_z. In numerical integration [2] this is called the "Trapezoidal Rule". Since the top boundary of the region is a series of straight line segments, it is easy to find these areas. The results are shown in Table 13.19. The columns in Table 13.19 are labeled L_{z1}, m_{z1}, R_{z1} to compare to those for $\overline{U}$ computed in the next section, and will be labeled L_{z2}, m_{z2}, R_{z2} and placed into Table 13.21.

Finally, we get to the objective function

$$\lambda_1 (1 - L_{z1i}) + \lambda_2 m_{z1i} + \lambda_3 R_{z1i}, \tag{13.30}$$

to be minimized. This is equation (13.19) with $K_1 = 1$, we may use this value of K_1 since all the values of L_{z1i} in Table 13.19 are between zero and one. Given values of the λ_i, substitute the L_{z1i}, m_{z1i} and R_{z1i} from Table 13.19, we find the value of i (Case number in Table 13.14) that makes the expression in equation (13.30) a minimum. The results for three choices for the λ_i values is shown in Table 13.20. The (a, b, c) labels at the top of this table means $\lambda_1 = a$, $\lambda_2 = b$ and $\lambda_3 = c$.

Case	$(1/3, 1/3, 1/3)$	$(0.4, 0.4, 0.2)$	$(0.3, 0.5, 0.2)$
1	0.5046	0.5451	0.4911
2	0.6607	0.6467	0.6157
3	0.4201	0.4752	0.4001
4	0.4288	0.4806	0.4072
5	0.4565	0.5076	0.4438
6	0.5409	0.5618	0.5089
7	0.4030	0.4617	0.3824
8	0.4055	0.4643	0.3849

Table 13.20: The Results of $min\overline{R}$ for the Eight Cases in Example 13.3.1.1

We see from Table 13.20 that the optimal solution is $c = 2, M = 4$, $\overline{\mu} = \overline{\mu}_2$ (Case 7) for all values of the λ_i chosen. However, there is another case, namely Case 8, whose values are very close to those of Case 7. So, Case 8 with $c = 2$, $M = 10$ and $\overline{\mu} = \overline{\mu}_2$, would be an alternate solution for all values of λ used.

13.3.2 Minimize $\overline{R}$ and Maximize $\overline{U}$

Let $\overline{Z1} = \overline{R}(M, c, \overline{\mu})$ and $\overline{Z2} = \overline{U}(M, c, \overline{\mu})$. Let the computations discussed above produce L_{zi}, m_{zi} and R_{zi} where we use $i = 1$ for $\overline{Z1}$ and $i = 2$ for $\overline{Z2}$.

We wish to maximize $\overline{U}$ which means minimize L_{z2}, maximize m_{z2} and maximize R_{z2}. We will turn these maximization goals around to be minimization objectives to be more compatible with minimize $\overline{R}$. Let K_2 (K_3) be a positive constant so that max m_{z2} (max R_{z2}) is equivalent to min $(K_2 - m_{z2})$ $(\min(K_3 - R_{z2}))$. Then for $\tau_i > 0$, $\tau_1 + \tau_2 + \tau_3 = 1$, in place of $max\overline{U}$ we use minimize

$$V2 = \tau_1 L_{z2} + \tau_2(K_2 - m_{z2}) + \tau_3(K_3 - R_{z2}). \tag{13.31}$$

Also let, from equation (13.19), us minimize

$$V1 = \lambda_1(K_1 - L_{z1}) + \lambda_2 m_{z1} + \lambda_3 R_{z1}. \tag{13.32}$$

First the decision maker chooses μ_1 and μ_2 for the two primary goals of $minV1$ and $minV2$ with $\mu_i > 0$, $\mu_1 + \mu_2 = 1$. We start with

$$min[\mu_1(V1) + \mu_2(V2)]. \tag{13.33}$$

Then we choose $\lambda_i > 0$, $\lambda_1 + \lambda_2 + \lambda_3 = 1$ and $\tau_i > 0$, $\tau_1 + \tau_2 + \tau_3 = 1$, and substitute $V1$ and $V2$ into equation (13.33) to obtain our overall objective function to minimize. The constraints on the variables are those in equation (13.20). Of course, we would vary the μ_i, λ_i and the τ_i to obtain a collection of solutions to discuss with management.

Case	L_{z2}	m_{z2}	R_{z2}
1	0.1020	0.6161	0.0948
2	0.1217	0.6628	0.1290
3	0.0468	0.1615	0.0758
4	0.0299	0.1615	0.0775
5	0.0880	0.5436	0.0899
6	0.1014	0.5705	0.1172
7	0.0354	0.1244	0.0500
8	0.0316	0.1270	0.0522

Table 13.21: Central Value and Certain Areas Under the Graph of $\overline{U}_i$, the Eight Cases in Example 13.3.2.1

Example 13.3.2.1

We consider the same eight cases in Table 13.14 with the same two values of the server rate $\overline{\mu}$. The objective is given in equation (13.33) with $V1$ from equation (13.32) and $V2$ in equation (13.31). In Example 13.3.1.1 we calculated L_{z1} and R_{z1} for $\overline{R}$ in the eight cases, so now we need to do the same for $\overline{U}$. We found the alpha-cuts of $\overline{U}$ for all the cases in Example 13.3.1.1 and now it is easy to compute L_{z2} and R_{z2}. We employ the same method, the trapezoidal rule, as explained in Example 13.3.1.1. The results are shown in Table 13.21.

Once we have these areas, it is easy to calculate the value of the objective function (see equation (13.33)) for various values of the parameters. Selections of the μ_i, λ_i and τ_i giving six cases are shown in Table 13.22. The cases I,II,...,VI across the top of the table are: (1) I is $\mu_1 = 0.6, \mu_2 = 0.4, \lambda_i = \tau_i = 1/3$ all i; (2) II equals $\mu_1 = 0.6, \mu_2 = 0.4, \lambda_1 = \lambda_2 = \tau_2 = \tau_3 = 0.4, \lambda_3 = \tau_1 = 0.2$; (3)case III has $\mu_1 = 0.6, \mu_2 = 0.4, \lambda_1 = \tau_3 = 0.3, \lambda_2 = \tau_2 = 0.5, \lambda_3 = \tau_1 = 0.2$; (4) IV is case I with $\mu_1 = 0.4, \mu_2 = 0.6$; (5) V is case II with $\mu = 0.4, \mu_2 = 0.6$; and (6) VI has the same as case III except $\mu_1 = 0.4, \mu_2 = 0.6$. In equation (13.31) for $V2$ we used $K_2 = K_3 = 1$ since m_{z2} and R_{z2} in Table 13.21 are all between zero and one.

Being a minimization problem we see from Table 13.22 that Case 5 ($c = 1, M = 4, \overline{\mu} = \overline{\mu}_2$) gives the optimal solution in situations I-IV, but Case 1 ($c = 1, M = 4, \overline{\mu} = \overline{\mu}_1$) is best for V and VI. However, an alternate solution in V and VI is Case 5.

13.3.3   Ranking the Fuzzy Sets

We discussed ranking fuzzy numbers in Section 2.6 of Chapter 2. Let us apply this to $min\overline{R}$ and to $max\overline{U}$. We first consider only $min\overline{R}$. Recall that the highest ranked fuzzy numbers are in the set H_K and the lowest ranked

Case	I	II	III	IV	V	VI
1	0.5125	0.5566	0.4882	0.4962	0.5624	0.4868
2	0.5737	0.6117	0.5511	0.5303	0.5942	0.5188
3	0.4933	0.5830	0.5224	0.5299	0.6369	0.5836
4	0.4961	0.5849	0.5251	0.5297	0.6371	0.5841
5	0.4678	0.5446	0.4738	0.4735	0.5632	0.4888
6	0.5130	0.5739	0.5053	0.4991	0.5800	0.5035
7	0.4899	0.5800	0.5214	0.5334	0.6391	0.5909
8	0.4903	0.5808	0.5218	0.5327	0.6390	0.5902

Table 13.22: Final Results of $min\overline{R}$ and $max\overline{U}$ in Example 13.3.2.1

fuzzy numbers are in H_1.

Fuzzy sets $\overline{A}_i$ are the values of $\overline{R}$ for $M = 4, 5, ..., 10$, $c = 1, 2, 3$ and $\overline{\mu} = \overline{\mu}_i$, $i = 1, 2$. Since we wish to minimize $\overline{R}$ we look at all the fuzzy sets in H_1. A problem may arise when we get more than one $\overline{R}$ in H_1. In general we suggest presenting management with all solutions in H_1 and let them pick one to implement. If we require a unique solution, then we could go back to the previous section to resolve the conflict.

Example 13.3.3.1

We will only use the $\alpha = 0$ and $\alpha = 1$ cuts of $\overline{R}$ and $\overline{U}$ in this section. We do have the $\alpha = 1/3$ and $\alpha = 2/3$ cuts of these fuzzy numbers but using only the $\alpha = 0$ and $\alpha = 1$ cuts will allow us to approximate $\overline{R}$ and $\overline{U}$ with triangular fuzzy numbers for ease in calculations and graphing. We did this also in Section 13.2.3. Using the triangular fuzzy number approximations will still give us a rough idea on how they line up from smallest to largest.

We first round off the $\alpha = 0$ cut and the $\alpha = 1$ cut of $\overline{R}_i$, $1 \leq i \leq 8$, to two decimal places for ease in producing a graph of these fuzzy numbers. Since we only use the $\alpha = 0$ and $\alpha = 1$ cut for these fuzzy numbers we will approximate them by triangular fuzzy numbers. For example, if the alpha equal zero cut of a $\overline{R}$ is $[r_1, r_3]$ and the alpha equal one cut is r_2, then $\overline{R} \approx (r_1/r_2/r_3)$. We will now drop the $\approx$ and write a $\overline{R}$ as a triangular fuzzy number. These triangular fuzzy numbers for all the $\overline{R}$ are shown in Table 13.23.

Next we graph these triangular fuzzy number from Table 13.23 to see how they line up from smallest to largest. This is now shown in Figure 13.3. In Figure 13.3 we only show the $\overline{R}_i$ in the lowest ranked set H_1, defined below. All these fuzzy numbers are very close together, and they are labeled, from left to right, as $\overline{R}_7$, $\overline{R}_8$, $\overline{R}_3$ and $\overline{R}_4$. We want H_1 the set of smallest fuzzy numbers. We will need to compute $v(\overline{R}_i, \overline{R}_j)$, see Section 2.6 in Chapter 2,

$\overline{R}$	Triangular Fuzzy Number
$\overline{R}_1$	(0.12/0.34/0.95)
$\overline{R}_2$	(0.12/0.47/2.34)
$\overline{R}_3$	(0.07/0.18/0.53)
$\overline{R}_4$	(0.06/0.19/0.62)
$\overline{R}_5$	(0.10/0.26/0.74)
$\overline{R}_6$	(0.09/0.33/1.46)
$\overline{R}_7$	(0,06/0.15/0.41)
$\overline{R}_8$	(0.06/0.16/0.39)

Table 13.23: The Triangular Fuzzy Numbers for the $\overline{R}$ in Example 13.3.3.1

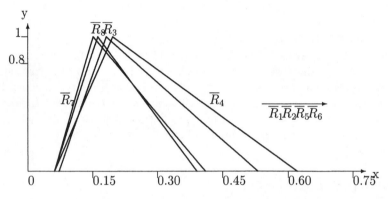

Figure 13.3: Ranking the Fuzzy Numbers $\overline{R}$ in Example 13.3.3.1

for $i \neq j \in \{1,2,3,4,5,6,7,8\}$. We will use $\eta = 0.8$ in the comparison as discussed in Section 2.6. We found that $H_1 = \{\overline{R}_3, \overline{R}_4, \overline{R}_7, \overline{R}_8\}$, $H_2 = \{\overline{R}_5\}$ and $H_3 = \{\overline{R}_1, \overline{R}_2, \overline{R}_6\}$. Notice that the height of the intersections between $\overline{R}_7$, $\overline{R}_8$, $\overline{R}_3$ and $\overline{R}_4$ all lie above the horizontal line at 0.8 on the $y-$axis, so they are all approximately equal to each other. If we choose $\overline{M}$ and $\overline{N}$ in H_1, then $\overline{M} \approx \overline{N}$, $\overline{M} \leq \overline{R}_5$ and $\overline{N} \leq \overline{R}_5$ with $\overline{R}_5 \in H_2$. Now $\overline{M} \leq \overline{R}_5$ means $\overline{M} < \overline{R}_5$ or $\overline{M} \approx \overline{R}_5$. So we could have $\overline{M} \approx \overline{R}_5$ for some $\overline{M} \in H_1$ but there will be some $\overline{Q} \in H_1$ so that $\overline{Q} < \overline{R}_5$. If we choose $\overline{O}$ and $\overline{P}$ in H_3, then $\overline{O} \approx \overline{P}$, $\overline{R}_5 \leq \overline{O}$ and $\overline{R}_5 \leq \overline{P}$. Again, there will be some $\overline{S} \in H_3$ so that $\overline{R}_5 < \overline{S}$.

Hence, the optimal solution consists of Cases 3,4,7 and 8. If we require a unique solution, then we may go back and use the analytical methods in Example 13.3.1.1 where Case 7 was the optimal solution.

Next consider maximizing $\overline{U}$ and minimizing $\overline{R}$. We obtain fuzzy sets $\overline{A}_{i1}$ for $\overline{R}$ and $\overline{A}_{i2}$ for $\overline{U}$. Find H_K for $\overline{U}$ and H_1 for $\overline{R}$. Hopefully there is a $\overline{A}_{i1} \in H_1$ for $M = M^*$, $c = c^*$ and $\overline{\mu} = \overline{\mu}^*$ and a $\overline{A}_{i2} \in H_K$ for the same

$\overline{U}$	Triangular Fuzzy Number
$\overline{U}_1$	(0.42/0.62/0.80)
$\overline{U}_2$	(0.43/0.66/0.91)
$\overline{U}_3$	(0.07/0.16/0.30)
$\overline{U}_4$	(0.12/0.16/0.32)
$\overline{U}_5$	(0.37/0.54/0.72)
$\overline{U}_6$	(0.38/0.57/0.81)
$\overline{U}_7$	(0.06/0.12/0.23)
$\overline{U}_8$	(0.06/0.13/0.23)

Table 13.24: The Triangular Fuzzy Numbers for the $\overline{U}$ in Example 13.3.3.2

values of the variables. Then these values of the variables is optimal. There may be other values of the variables that are optimal, then we may have to resolve the conflict for a unique solution. However, there may be no optimal values of the variables and then we abandon this procedure and go to Section 13.3.2.

Example 13.3.3.2

We first round off the $\alpha = 0$ cut and the $\alpha = 1$ cut of $\overline{U}_i$, $1 \leq i \leq 8$, to two decimal places for ease in producing a graph of these fuzzy numbers. Since we only use the $\alpha = 0$ and $\alpha = 1$ cut for these fuzzy numbers we will approximate them by triangular fuzzy numbers. For example, if the alpha equal zero cut of a $\overline{U}$ is $[u_1, u_3]$ and the alpha equal one cut is u_2, then $\overline{U} \approx (u_1/u_2/u_3)$. We will now drop the $\approx$ and write a $\overline{U}$ as a triangular fuzzy number. These triangular fuzzy numbers for all the $\overline{U}$ are shown in Table 13.24.

Next we graph these triangular fuzzy numbers from Table 13.24 to see how they line up from smallest to largest. This is now shown in Figure 13.4. In Figure 13.4 we only show the $\overline{U}_i$ in the highest ranked set H_4, defined below. We want H_K the set of largest fuzzy numbers. We will need to compute $v(\overline{U}_i, \overline{U}_j)$, see Section 2.6 in Chapter 2, for $i \neq j \in \{1, 2, 3, 4, 5, 6, 7, 8\}$. We will use $\eta = 0.8$ in the comparison as discussed in Section 2.6. We found that $H_1 = \{\overline{U}_3, \overline{U}_7, \overline{U}_8\}$, $H_2 = \{\overline{U}_4\}$, $H_3 = \{\overline{U}_5\}$ and $H_4 = \{\overline{U}_1, \overline{U}_2, \overline{U}_6\}$. The height of the intersections between $\overline{U}_1$, $\overline{U}_2$ and $\overline{U}_6$ all are above the 0.8 level on the vertical axis, so they are all approximately equal to each other. If we choose $\overline{M}$ and $\overline{N}$ in H_1, then $\overline{M} \approx \overline{N}$, $\overline{M} \leq \overline{U}_4$ and $\overline{N} \leq \overline{U}_4$ with $\overline{U}_4 \in H_2$. Now $\overline{M} \leq \overline{U}_4$ means $\overline{M} < \overline{U}_4$ or $\overline{M} \approx \overline{U}_4$. So we could have $\overline{M} \approx \overline{U}_4$ for some $\overline{M} \in H_1$ but there will be some $\overline{Q} \in H_1$ so that $\overline{Q} < \overline{U}_4$. Similar statements can be made when comparing H_i and H_j for $i < j$. Hence, the optimal solution consists of Cases 1,2,and 6.

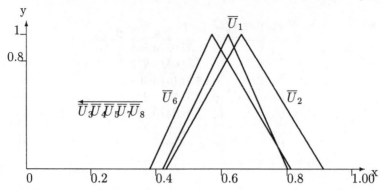

Figure 13.4: Ranking the Fuzzy Numbers $\overline{U}$ in Example 13.3.3.2

Comparing the results on $min\overline{R}$ and $max\overline{U}$ we have a conflict. For $min\overline{R}$ the optimal cases were 3,4,7,8 and for $max\overline{U}$ the optimal cases are 1,2,6. We definitely have conflicting goals. To resolve this conflict we may go back and employ the analytical methods in Example 13.3.2.1.

13.4   References

1. J.J.Buckley: Fuzzy Probabilities: New Approach and Applications, Physica-Verlag, Heidelberg, 2003.

2. R.L.Finney and G.B.Thomas: Calculus, Second Edition, Addison-Wesley, New York, 1994.

3. Frontline Systems (www.frontsys.com).

4. Maple 6, Waterloo Maple Inc., Waterloo, Canada.

5. H.A.Taha: Operations Research, Fifth Edition, Macmillan, N.Y., 1992.

Chapter 14

Optimization: With Revenue/Costs

14.1 Introduction

In the previous Chapter we looked at optimizing the system without considering costs and revenues. Now we will construct an optimizing model incorporating these factors. Again we have two models. The first one uses fuzzy probabilities and the second model employs fuzzy arrival/sevice rates.

14.2 Fuzzy Probabilities

There are two types of servers, one with associated fuzzy probability $\overline{p}_1$ and the other faster server with fuzzy probability $\overline{p}_2$. Let $\overline{C}_i$ be the cost, in \$ per unit time, of operating server having corresponding fuzzy probability $\overline{p}_i$, $i = 1, 2$. Determining the cost of a server per unit time is a difficult number to estimate so we assign a fuzzy number $\overline{C}_i$ to its value.

Some of the fuzzy probabilities and constants in our model may have to be estimated by experts. We discussed this problem in Section 3.4. So we assume this has been accomplished and we have the fuzzy numbers for these parameters.

There is a cost involved in maintaining the queue, or those in the system but not yet in a server. Let fuzzy $\overline{Q}$ be the cost, in \$ per unit time, of having one space available in the queue. So the queue cost is $\overline{Q}(M - c)$.

There will be certain fixed costs associated with maintaining the web site which do not depend on the decision variables. Since they do not depend on $M, c,...$ they can be omitted for the model.

We will assume that we can effect the fuzzy probabilities $\overline{p}(i)$, the fuzzy probability that i customers arrive at the system per unit time, through advertising. Let $\overline{A}_1$ be the advertising level per unit time producing $\overline{p}^{(1)}(i)$, $0 \leq i \leq L_1$, and $\overline{A}_2$ the advertising amount in \$ per unit time that is expected to produce $\overline{p}^{(2)}(i)$, $0 \leq i \leq L_2$. $\overline{A}_2$ was designed to increase arrivals per unit time so $\overline{p}^{(2)}(i)$ is $\overline{p}^{(1)}(i)$ shifted slightly to the right. We have assumed that $\overline{p}^{(u)}(i) = 0$ for $i > L_u$, $u = 1, 2$.

Revenue from advertisers is assumed to be proportional to the average number of customers in the system $\overline{N}$. If $\overline{T}$ is the revenue, in \$ per unit time, per customer in the system, then total revenue per unit time is $\overline{T}\,\overline{N}$.

Fuzzy profit per unit time, to be maximized, is

$$\overline{Profit} = \overline{T}\,\overline{N} - [\overline{A}_u + \overline{Q}(M - c) + \overline{C}_v c], \qquad (14.1)$$

where: (1) we use $\overline{p}^{(u)}(i)$ with $\overline{A}_u$, $u = 1, 2$ in computing $\overline{N}$; and (2) we use $\overline{p} = \overline{p}_v$ with $\overline{C}_v$, $v = 1, 2$, for $\overline{N}$. We modeled all the cost/income parameters as fuzzy numbers, if any are known exactly, then we would use their exact values. The other variables are $c = 1, 2, 3, 4$ and $M = c, c + 1, ..., maxM$. We could consider a budget constraint, only so much money available per unit time, but we will not do this here. The above equation gives us one goal. We could add other goals such as $min\overline{R}$, $max\overline{U}$ and $min\overline{LC}$.

There are many other costs associated with the system, such as startup costs and operating costs ([1], Chapter 5), which we have not incorporated into the model. Many of these costs are independent from our variables, so can be classified as fixed costs in our model, and hence omitted. However, some other costs could be included to make our optimization model more complete.

Example 14.2.1

We first present our strategy for maximizing fuzzy profit and then we discuss ranking the fuzzy sets for profit from smallest to largest.

$max\overline{Profit}$

We concentrate on the sixteen cases in Table 14.1. We will be using for the "normal" fuzzy arrival probabilities $\overline{p}^{(1)}(i)$ those given in Table 6.1 and $\overline{p}_1 = (0.3/0.4/0.5)$, $\overline{p}_2 = (0.5/0.6/0.7)$ for the two fuzzy service rates. These were all used together in Examples 13.2.1.1 and 13.2.2.1 in Chapter 13. For our "faster" fuzzy arrival probabilities $\overline{p}^{(2)}(i)$, whose values we obtain from expert opinion, we will employ those shown in Table 14.2. Notice that $\overline{p}^{(2)}(i) = 0$ for $i \geq 5$.

We also need fuzzy numbers for the constants in equation (14.1). Through data analysis, and some expert opinion, we come up with the triangular fuzzy numbers for all these parameters which are shown in Table 14.3.

Case	Number of Servers	System Capacity	Server Rate	Arrival Rate
1	$c = 1$	$M = 4$	$\overline{p}_1$	$\overline{p}^{(1)}(i)$
2	$c = 1$	$M = 10$	$\overline{p}_1$	$\overline{p}^{(1)}(i)$
3	$c = 2$	$M = 4$	$\overline{p}_1$	$\overline{p}^{(1)}(i)$
4	$c = 2$	$M = 10$	$\overline{p}_1$	$\overline{p}^{(1)}(i)$
5	$c = 1$	$M = 4$	$\overline{p}_2$	$\overline{p}^{(1)}(i)$
6	$c = 1$	$M = 10$	$\overline{p}_2$	$\overline{p}^{(1)}(i)$
7	$c = 2$	$M = 4$	$\overline{p}_2$	$\overline{p}^{(1)}(i)$
8	$c = 2$	$M = 10$	$\overline{p}_2$	$\overline{p}^{(1)}(i)$
9	$c = 1$	$M = 4$	$\overline{p}_1$	$\overline{p}^{(2)}(i)$
10	$c = 1$	$M = 10$	$\overline{p}_1$	$\overline{p}^{(2)}(i)$
11	$c = 2$	$M = 4$	$\overline{p}_1$	$\overline{p}^{(2)}(i)$
12	$c = 2$	$M = 10$	$\overline{p}_1$	$\overline{p}^{(2)}(i)$
13	$c = 1$	$M = 4$	$\overline{p}_2$	$\overline{p}^{(2)}(i)$
14	$c = 1$	$M = 10$	$\overline{p}_2$	$\overline{p}^{(2)}(i)$
15	$c = 2$	$M = 4$	$\overline{p}_2$	$\overline{p}^{(2)}(i)$
16	$c = 2$	$M = 10$	$\overline{p}_2$	$\overline{p}^{(2)}(i)$

Table 14.1: The Sixteen Cases in Example 14.2.1

$\overline{p}$	Fuzzy Probability
$\overline{p}(0)$	0
$\overline{p}(1)$	$(0.3/0.4/0.5)$
$\overline{p}(2)$	$(0.2/0.3/0.4)$
$\overline{p}(3)$	$(0.1/0.2/0.3)$
$\overline{p}(4)$	$(0.0/0.1/0.2)$
$\overline{p}(5)$	0
$\overline{p}(6)$	0
$\overline{p}(7)$	0

Table 14.2: Fuzzy Probabilities for Arrivals $\overline{p}^{(2)}(i)$ in Example 14.2.1

Constant	Fuzzy Number
$\overline{A}_1$	$(0.09/0.15/0.21)$
$\overline{A}_2$	$(0.21/0.27/0.33)$
$\overline{C}_1$	$(0.27/0.30/0.33)$
$\overline{C}_2$	$(028/0.34/0.40)$
$\overline{Q}$	$(0.04/0.07/0.10)$
$\overline{T}$	$(0.83/0.95/1.07)$

Table 14.3: Fuzzy Numbers for the Constants in Equation (14.1) for Example 14.2.1

As in Example 13.2.1.1 we will only compute the $\alpha = 0$ and $\alpha = 1$ cuts of the fuzzy steady state probabilities. This means that we will only have the $\alpha = 0$ and $\alpha = 1$ cuts of $\overline{Profit}$. We will number the fuzzy numbers $\overline{N}_i$ according to the sixteen cases $i = 1, 2, 3, ..., 16$ in Table 14.1. Now we go through four steps: (1) step 1 is to find the fuzzy $\overline{p}_{ij}$, only for $\alpha = 0, 1$, in the fuzzy transition matrices $\overline{P}$; (2) calculate the fuzzy steady state probabilities, for $\alpha = 0, 1$, in Step 2 ; (3) Step 3 determines the $\overline{N}_i$, $1 \leq i \leq 16$, for alpha zero and one; and (4) Step 4 solves the maximization of fuzzy profit with fuzzy profit given in equation (14.1).

Step 1

This was accomplished for Cases 1 through 8 in Example 13.2.1.1 in Chapter 13. Here we need to do this for Cases 9 through 16 using the new fuzzy arrival probabilities in Table 14.2. However, the calculations now are a bit easier because the transition matrix will have more zeros in it. From Table 14.2 we know that $p(0) = 0$ and $p(i) = 0$ for $i \geq 5$. Therefore, from Table 6.2, in Cases 9 and 13 we have $p_{00} = p_{10} = p_{20} = p_{21} = p_{30} = p_{31} = p_{32} = p_{40} = p_{41} = p_{42} = p_{43} = 0$, $p_{44} = 1$. From Table 8.1 for Cases 11 and 15 we see $p_{00} = p_{10} = p_{20} = p_{30} = p_{31} = p_{40} = p_{41} = p_{42} = 0$. Next we describe the 11×11 crisp transition matrix P that goes with Cases 10 and 14, where $c = 1$ and $M = 10$, as we did in Example 13.2.1.1. The rows/columns are numbered $0, 1, 2, ..., 10$.

1. row #0: 0,p(1),p(2),p(3),p(4),0,0,0,0,0,0;

2. row #1: 0,b,c,d,e,f,0,0,0,0,0;

3. row #2: 0,0,b,c,d,e,f,0,0,0,0;

4. row #3: 0,0,0,b,c,d,e,f,0,0,0;

5. row #4: 0,0,0,0,b,c,d,e,f,0,0;

6. row #5: 0,0,0,0,0,b,c,d,e,f,0;

7. row #6: 0,0,0,0,0,0,b,c,d,e,f^*;

8. row #7: 0,0,0,0,0,0,0,b,c,d,e^*;

9. row #8: 0,0,0,0,0,0,0,0,b,c,d^*;

10. row #9: 0,0,0,0,0,0,0,0,0,b,c^*;

11. row #10: 0,0,0,0,0,0,0,0,0,0,1;

where

1. $b = p(1)p$;

2. $c = p(2)p + p(1)(1 - p)$;

3. $d = p(3)p + p(2)(1 - p)$;

4. $e = p(4)p + p(3)(1 - p)$;

5. $f = p(4)(1 - p)$;

6. $f^* = f$;

7. $e^* = p^*(3)(1 - p) + p(4)p$;

8. $d^* = p^*(2)(1 - p) + p^*(3)p$;

9. $c^* = p^*(1)(1 - p) + p^*(2)p$;

10. $b^* = p^*(1)p$;

where $p^*(i) = p(i) + p(i + 1) + ... + p(4)$.

The 11×11 crisp transition matrix P for Cases 12 and 16 , for $c = 2$ and $M = 10$, is

1. row #0: 0,p(1),p(2),p(3),p(4),0,0,0,0,0,0;

2. row #1: 0,b,c,d,e,f,0,0,0,0,0;

3. row #2: 0,B,C,D,E,F,G,0,0,0,0;

4. row #3: 0,0,B,C,D,E,F,G,0,0,0;

5. row #4: 0,0,0,B,C,D,E,F,G,0,0;

6. row #5: 0,0,0,0,B,C,D,E,F,G,0;

7. row #6: 0,0,0,0,0,B,C,D,E,F,G^*;

8. row #7: 0,0,0,0,0,0,B,C,D,E,F^*;

9. row #8: 0,0,0,0,0,0,0,B,C,D,E^*;

10. row #9: 0,0,0,0,0,0,0,0,B,C,D^*;

11. row #10: 0,0,0,0,0,0,0,0,0,B,C^*;

where

1. b,...,f were all defined above for $c = 1$;

2. $B = p(1)q(2|2)$;

3. $C = p(1)q(1|2) + p(2)q(2|2)$;

4. $D = p(1)q(0|2) + p(2)q(1|2) + p(3)q(2|2)$;

5. $E = p(2)q(0|2) + p(3)q(1|2) + p(4)q(2|2)$;

6. $F = p(3)q(0|2) + p(4)q(1|2)$;

7. $G = p(4)q(0|2)$;

8. $G^* = G$;

9. $F^* = p^*(3)q(0|2) + p(4)q(1|2)$;

10. $E^* = p^*(2)q(0|2) + p^*(3)q(1|2) + p(4)q(2|2)$;

11. $D^* = p^*(1)q(0|2) + p^*(2)q(1|2) + p^*(3)q(2|2)$;

12. $C^* = q(0|2) + p^*(1)q(1|2) + p^*(2)q(2|2)$;

where $p^*(i) = p(i) + p(i + 1) = ... + p(4)$.

In general, all the non-zero $\alpha = 0$ cuts of the $\overline{p}_{ij}$ in the fuzzy transition matrix $\overline{P}$ were calculated as before in Example 13.2.1.1 using $\overline{p}_1$ or $\overline{p}_2$

We now assume that all the alpha equal zero cuts have been found for all the elements in all the fuzzy transition matrices in Cases 9 through 16.

In Cases 9,10,13 and 14 the fuzzy transition matrices are fuzzy transition matrices for a fuzzy, absorbing, Markov chain discussed in Section 4.3 of Chapter 4. It follows that the fuzzy steady state probabilities are all crisp numbers. See Example 4.3.1, and the discussion following this example, because there will be only one absorbing state. In these cases the fuzzy steady state probabilities are: (1) for $M = 4$ we get $\overline{w} = (0, 0, 0, 0, 1)$ or $\overline{w}_4 = 1$; and (2) for $M = 10$ we find $\overline{w} = (0, 0, 0, 0, 0, 0, 0, 0, 0, 0, 1)$ or $\overline{w}_{10} = 1$.

In Cases 11,12,15 and 16 the fuzzy transition matrices are fuzzy transition matrices for the other type of fuzzy Markov chain discussed in Section 4.4 of Chapter 4. The fuzzy steady state probabilities will be: (1) for $M = 4$, we have $\overline{w} = (0, \overline{w}_1, ..., \overline{w}_4)$; and (2) when $M = 10$ we obtain $\overline{w} = (0, \overline{w}_1, ..., \overline{w}_{10})$.

Step 2

Here we determine (estimate) the end points of the $\alpha = 0$ cut of the fuzzy steady state probabilities for Cases 11,12,15,16 in Table 14.1. The calculations for Cases 1-8 were completed in Example 13.2.1.1 and we know the results for Cases 9,10,13,14. We present only two tables: (1) Table 14.4 is for Case 11 and Table 14.5 has the alpha-cuts in Case 16 (0.0000* means it is 0.0000 rounded off to four decimal places).

The $\overline{w}_i[1]$ are easy to get since the $\alpha = 1$ cut of the fuzzy transition matrix is just a crisp matrix P, whose $p_{ij} \in [0, 1]$ and the row sums of P are all one. Therefore P^n converges as $n \to \infty$. The $\overline{w}_i[1]$ are in P^n for sufficiently large n.

$\overline{w}$	$\alpha = 0$	$\alpha = 1$
$\overline{w}_0[\alpha]$	$[0,0]$	0
$\overline{w}_1[\alpha]$	$[0.0000^*, 0.0063]$	0.0005
$\overline{w}_2[\alpha]$	$[0.0010, 0.0379]$	0.0070
$\overline{w}_3[\alpha]$	$[0.0308, 0.1756]$	0.0800
$\overline{w}_4[\alpha]$	$[0.7827, 0.9681]$	0.9125

Table 14.4: $\alpha = 0, 1$ Cuts of the Fuzzy Steady State Probabilities, Case 11 in Example 14.2.1

$\overline{w}$	$\alpha = 0$	$\alpha = 1$
$\overline{w}_0[\alpha]$	$[0,0]$	0
$\overline{w}_1[\alpha]$	$[0.0000^*, 0.0005]$	0.0000^*
$\overline{w}_2[\alpha]$	$[0.0000^*, 0.0021]$	0.0000^*
$\overline{w}_3[\alpha]$	$[0.0000^*, 0.0058]$	0.0000^*
$\overline{w}_4[\alpha]$	$[0.0000^*, 0.0129]$	0.0001
$\overline{w}_5[\alpha]$	$[0.0000^*, 0.0276]$	0.0005
$\overline{w}_6[\alpha]$	$[0.0001, 0.0492]$	0.0022
$\overline{w}_7[\alpha]$	$[0.0008, 0.0857]$	0.0094
$\overline{w}_8[\alpha]$	$[0.0086, 0.1524]$	0.0409
$\overline{w}_9[\alpha]$	$[0.0862, 0.3113]$	0.1773
$\overline{w}_{10}[\alpha]$	$[0.4802, 0.9007]$	0.7696

Table 14.5: $\alpha = 0, 1$ Cuts of the Fuzzy Steady State Probabilities, Case 16 in Example 14.2.1

$\overline{N}$	$\alpha = 0$	$\alpha = 1$
$\overline{N}_1$	[3.9037, 3.9752]	3.9476
$\overline{N}_2$	[9.9036, 9.9752]	9.9476
$\overline{N}_3$	[3.5735, 3.9129]	3.8083
$\overline{N}_4$	[9.5270, 9.9123]	9.7903
$\overline{N}_5$	[3.8473, 3.9192]	3.9114
$\overline{N}_6$	[9.8456, 9.9557]	9.9144
$\overline{N}_7$	[3.1994, 3.7954]	3.5801
$\overline{N}_8$	[9.0089, 9.7869]	9.5262
$\overline{N}_9$	[4, 4]	4
$\overline{N}_{10}$	[10, 10]	10
$\overline{N}_{11}$	[3.7322, 3.9671]	3.9045
$\overline{N}_{12}$	[9.7196, 9.9671]	9.9042
$\overline{N}_{13}$	[4, 4]	4
$\overline{N}_{14}$	[10, 10]	10
$\overline{N}_{15}$	[3.1811, 3.8936]	3.7186
$\overline{N}_{16}$	[8.7804, 9.8902]	9.7008

Table 14.6: $\alpha = 0, 1$ Cuts of the Fuzzy Expected Number in the System $\overline{N}$, the Sixteen Cases in Example 14.2.1

Step 3

Here we only need to compute $\overline{N}$ for equation (14.1) only for Cases 11,12,15,16 because $\overline{N} = 4(10)$ in Cases 9,10,13,14 when $M = 4(10)$. The $\alpha = 0$ cut of $\overline{N}$ is found by solving a linear programming problem as discussed for the $M = 4$ cases in Chapters 6-9. So let us only look at how to get $\overline{N}[0] = [n_1(0), n_2(0)]$ when $M = 10$. The linear programming problems to solve are

$$max/min[w_1 + 2w_2 + \ldots + 10w_{10}], \qquad (14.2)$$

subject to

$$w_{i1}(0) \leq w_i \leq w_{i2}(0), 0 \leq i \leq 10, w_0 + \ldots + w_{10} = 1, \qquad (14.3)$$

where $\overline{w}_i[0] = [w_{i1}(0), w_{i2}(0)]$, $0 \leq i \leq 10$. The max (min) solution produces $n_2(0)$ ($n_1(0)$). We used the program "simplex" within Maple [2] for these linear programming problems. Table 14.6 gives the $\alpha = 0, 1$ cuts of $\overline{N}$ for the sixteen cases in Table 14.1.

Step 4

We are almost ready to find the values of the variables to maximize fuzzy profit. First we will determine the alpha equal to zero cut of $\overline{Profit}$ in

Constant	$\alpha = 0$ Cut
$\overline{A}_1$	$[0.09, 0.21]$
$\overline{A}_2$	$[0.21, 0.33]$
$\overline{C}_1$	$[0.27, 0.33]$
$\overline{C}_2$	$[0.28, 0.40]$
$\overline{Q}$	$[0.04, 0.10]$
$\overline{T}$	$[0.83, 1.07]$
$\overline{N}$	$[n_1(0), n_2(0)]$

Table 14.7: $\alpha = 0$ Cuts of the Fuzzy Numbers in Equation (14.1) for Example 14.2.1

equation (14.1). We do this using interval arithmetic (see Section 2.3.2 of Chapter 2). Let us go through some of these details. We will need the $\alpha = 0$ cuts shown in Table 14.7.

Then

$$\overline{Profit}[0] = [0.83n_1(0), 1.07n_2(0)] - [s, t], \qquad (14.4)$$

where s and t depend on using $\overline{p}^{(u)}(i)$ for $u = 1, 2$ and on using $\overline{p}_v$ for $v = 1, 2$. To illustrate these calculations let us assume that we are employing $\overline{p}^{(2)}(i)$ and $\overline{p}_2$. Then

$$[s, t] = [0.21 + 0.04(M - c) + 0.28c, 0.33 + 0.10(M - c) + 0.40c], \qquad (14.5)$$

and

$$\overline{Profit}[0] = [0.83n_1(0) - t, 1.07n_2(0) - s]. \qquad (14.6)$$

Now $\overline{Profit}$ is a function of c, M, $\overline{p}^{(u)}$ and $\overline{p}_v$ and we show this as writing $\overline{Z} = \overline{Profit}(M, c, \overline{p}^{(u)}, \overline{p}_v)$. So, for different values of the variables we get triangular shaped fuzzy numbers $\overline{Z}$. Find the values of the variables to maximize $\overline{Z}$. We number fuzzy profit $\overline{Profit}_i$, $1 \leq i \leq 16$, for the sixteen cases in Table 14.1. Let: (1) m_z be the center of the core of $\overline{Z}$; (2) L_z be the area under the graph of the membership function to the left of m_z; and (3) R_z be the area under the graph of the membership function to the right of m_z. As before we finally obtain the objective function

$$max(\lambda_1[K_1 - L_z] + \lambda_2 m_z + \lambda_3 R_z), \qquad (14.7)$$

where $\lambda_i > 0$, $1 \leq i \leq 3$, $\lambda_1 + \lambda_2 + \lambda_3 = 1$, and the values of the variables are shown in Table 14.1. Of course we pick the constant K_1 sufficiently large so that $minL_z$ is equivalent to $max[K_1 - L_z]$.

What we need to do is compute L_z and R_z for $\overline{Profit}_i$, $i = 1, 2, ..., 16$. All the fuzzy numbers in equation (14.1) are triangular fuzzy numbers. Recall we only have the $\alpha = 0, 1$ cuts of $\overline{N}$ and we therefore approximate $\overline{N}$ by a triangular fuzzy number. It follows that $\overline{Profit}$ is also a triangular fuzzy

Case	L_z	m_z	R_z
1	0.3451	3.0902	0.3566
2	0.7951	8.3702	0.8066
3	0.4159	2.7279	0.3894
4	0.8767	7.9908	0.8477
5	0.3663	3.0158	0.3588
6	0.8184	8.2987	0.8270
7	0.4928	2.4311	0.4650
8	0.9962	7.6599	0.9210
9	0.3300	3.0200	0.3300
10	0.7800	8.3000	0.7800
11	0.3958	2.6993	0.3578
12	0.8509	7.9790	0.8079
13	0.3450	2.9800	0.3450
14	0.7950	8.2600	0.7950
15	0.5662	2.4427	0.4367
16	1.1740	7.7058	0.8934

Table 14.8: Central Value and Certain Areas Under the Graph of $\overline{Profit}_i$, for the Sixteen Cases in Example 14.2.1

number. The areas L_z and R_z become the areas of right triangles of height one and are easily found. The results are shown in Table 14.8.

Finally, we get to evaluate the objective function to be maximized. In equation (14.7) we will use $K_1 = 1$ even though there is a value of L_z in Table 14.8 that exceeds one (Case 16). The fact that the term $(K_1 - L_z)$ can be negative will not effect the optimal solution to the max problem in equation (14.7). Given values of the λ_i, substitute the L_z, m_z and R_z from Table 14.8, we find the value of i (Case number in Table 14.1) that makes the expression in equation (14.7) a maximum. The results for three choices for the λ values is shown in Table 14.9. The (a, b, c) labels at the top of this table means $\lambda_1 = a$, $\lambda_2 = b$ and $\lambda_3 = c$.

From Table 14.9 Case 2 is best for maximizing fuzzy profit for all values of the λ_i given. Case 2 is $c = 1$, $M = 10$, $\overline{p} = \overline{p}_1 = (0.3/0.4/0.5)$ the slower fuzzy service rate and $\overline{p}(i) = \overline{p}^{(1)}(i)$ the "slower" fuzzy arrival probabilities, employing $\overline{C} = \overline{C}_1$ and $\overline{A} = \overline{A}_1$ (see Table 14.3). There are a number of alternate solutions, whose value for $max\overline{Profit}$ is close to that of Case 2. In all cases the system is congested (see Table 14.6) since $\overline{N}$ is at its maximum M in cases 9,10,13, and 14 and it is close to its maximum in all other cases. So we will add the objective of minimizing lost customer in Chapter 17.

Case	$(1/3, 1/3, 1/3)$	$(0.2, 0.4, 0.4)$	$(0.2, 0.5, 0.3)$
1	1.3672	1.5097	1.7831
2	3.1272	3.7117	4.4681
3	1.2338	1.3637	1.5976
4	2.9873	3.5600	4.2744
5	1.3361	1.4766	1.7423
6	3.1024	3.6866	4.4337
7	1.1344	1.2599	1.4565
8	2.8616	3.4331	4.1070
9	1.3400	1.4740	1.7430
10	3.1000	3.6760	4.4280
11	1.2204	1.3436	1.5778
12	2.9787	3.5446	4.2617
13	1.3267	1.4610	1.7245
14	3.0867	3.6630	4.4095
15	1.1044	1.2385	1.4391
16	2.8084	3.4048	4.0861

Table 14.9: The Results of $max\overline{Profit}$ for the Sixteen Cases in Example 14.2.1

14.2.1 Ranking the Fuzzy Sets

We discussed ranking fuzzy numbers in Section 2.6 of Chapter 2. Let us apply this to $max\overline{\Pi}$. Recall that the highest ranked fuzzy numbers are in the set H_K and the lowest ranked fuzzy numbers are in H_1.

Fuzzy sets $\overline{A}_i$ are the values of $\overline{\Pi}$. Since we wish to maximize $\overline{\Pi}$ we look at all the fuzzy sets in H_K. A problem may arise when we get more than one $\overline{\Pi}$ in H_K. In general we suggest presenting management with all solutions in H_K and let them pick one to implement. If we require a unique solution, then we could go back to the previous section to resolve the conflict.

Example 14.2.1.1

We first round off the $\alpha = 0$ cut and the $\alpha = 1$ cut of $\overline{\Pi}$, $1 \le i \le 16$, to one decimal place for ease in producing a graph of these fuzzy numbers. Since we only have the $\alpha = 0$ and $\alpha = 1$ cut for these fuzzy numbers we will approximate them by triangular fuzzy numbers. For example, if the alpha equal zero cut of a $\overline{\Pi}$ is $[\pi_1, \pi_3]$ and the alpha equal one cut is π_2, then $\overline{\Pi} \approx (\pi_1/\pi_2/\pi_3)$. We will now drop the $\approx$ and write a $\overline{\Pi}$ as a triangular fuzzy number. These triangular fuzzy numbers for all the $\overline{\Pi}$ are shown in Table 14.10.

From Table 14.10 we see that the fuzzy numbers $\overline{\Pi}$ divide into two groups:

$\overline{\Pi}$	Triangular Fuzzy Number
$\overline{\Pi}_1$	(2.4/3.1/3.8))
$\overline{\Pi}_2$	(6.8/8.4/10.0)
$\overline{\Pi}_3$	(1.9/2.7/3.5)
$\overline{\Pi}_4$	(6.2/8.0/9.7)
$\overline{\Pi}_5$	(2.3/3.0/3.7)
$\overline{\Pi}_6$	(6.7/8.3/10.0)
$\overline{\Pi}_7$	(1.4/2.4/3.4)
$\overline{\Pi}_8$	(5.7/7.7/9.5)
$\overline{\Pi}_9$	(2.4/3.0/3.7)
$\overline{\Pi}_{10}$	(6.7/8.3/9.9)
$\overline{\Pi}_{11}$	(1.9/2.7/3.4)
$\overline{\Pi}_{12}$	(6.3/8.0/9.6)
$\overline{\Pi}_{13}$	(2.3/3.0/3.7)
$\overline{\Pi}_{14}$	(6.7/8.3/9.8)
$\overline{\Pi}_{15}$	(1.3/2.4/3.3)
$\overline{\Pi}_{16}$	(5.4/7.7/9.5)

Table 14.10: The Triangular Fuzzy Numbers for the $\overline{\Pi}$ in Example 14.2.1.1

(1) $G_1 = \{\overline{\Pi}_1, \overline{\Pi}_3, \overline{\Pi}_5, \overline{\Pi}_7, \overline{\Pi}_9, \overline{\Pi}_{11}, \overline{\Pi}_{13}, \overline{\Pi}_{15}\}$ the smaller fuzzy numbers; and (2) $G_2 = $ the rest of the fuzzy numbers being the larger fuzzy numbers. We obtain $\overline{\Pi}_i < \overline{\Pi}_j$ for $i = 1, 3, 5, 7, 9, 11, 13, 15$ and $j = 2, 4, 6, 8, 10, 12, 14, 16$. To find the highest ranked $\overline{\Pi}_i$ we need to study group G_2. We will not show the graph of all the fuzzy number in G_2 because there are eight of them and they are all bunched together and it would be difficult to identify any one of them (see Figure 13.3 for just four fuzzy numbers). Instead we evaluated $v(\overline{\Pi}_i, \overline{\Pi}_j)$, see Section 2.6 in Chapter 2, for $i \neq j \in \{2, 4, 6, 8, 10, 12, 14, 16\}$. In all of these cases we computed a value between 0.80 and 1.00. We are using $\eta = 0.80$ in Section 2.6 in Chapter 2. Hence, the highest ranked fuzzy numbers $\overline{\Pi}_i$ are those in G_2 where $\overline{\Pi}_i \approx \overline{\Pi}_j$ for $i \neq j \in \{2, 4, 6, 8, 10, 12, 14, 16\}$. The optimal solution is Cases 2,4,6,8,10,12,14,16. To get a unique optimal solution we may go back to the analytical method shown in Example 14.2.1. where Case 2 was the optimal solution.

14.3   Fuzzy Arrival/Service Rates

There are two types of servers, one with associated fuzzy service rate $\overline{\mu}_1$ and the other faster server with fuzzy service rate $\overline{\mu}_2$. Let $\overline{C}_i$ be the cost, in \$ per unit time, of operating server having corresponding fuzzy service rate $\overline{\mu}_i$, $i = 1, 2$. Determining the cost of a server per unit time is a difficult number to estimate so we assign a fuzzy number $\overline{C}_i$ to its value.

There is a cost involved in maintaining the queue, or those in the system but not yet in a server. Let fuzzy $\overline{Q}$ be the cost, in \$ per unit time, of having one space available in the queue. So the queue cost is $\overline{Q}(M - c)$.

There will be certain fixed costs associated with maintaining the web site which do not depend on the decision variables. Since they do not depend on $M, c,...$ they can be omitted for the model.

We will assume that we can effect the fuzzy arrival rate $\overline{\lambda}$ through advertising. Let $\overline{A}_1$ be the advertising level per unit time producing $\overline{\lambda}_1$ and $\overline{A}_2$ the advertising amount in \$ per unit time that is expected to produce $\overline{\lambda}_2$. $\overline{A}_2$ was designed to increase arrivals per unit time so $\overline{\lambda}_2$ is $\overline{\lambda}_1$ shifted to the right.

Revenue from advertisers is assumed to be proportional to the average number of customers in the system $\overline{N}$. If $\overline{T}$ is the revenue, in \$ per unit time, per customer in the system, then total revenue per unit time is $\overline{T}\,\overline{N}$.

Fuzzy profit per unit time, to be maximized, is

$$\overline{Profit} = \overline{T}\,\overline{N} - [\overline{A}_u + \overline{Q}(M - c) + \overline{C}_v c], \qquad (14.8)$$

where: (1) we use $\overline{\lambda}_u$ with $\overline{A}_u$, $u = 1, 2$ in computing $\overline{N}$; and (2) we use $\overline{\mu} = \overline{\mu}_v$ with $\overline{C}_v$, $v = 1, 2$, for $\overline{N}$. We modeled all the cost/income parameters as fuzzy numbers, if any are known exactly, then we would use their exact values. The other variables are $c = 1, 2, 3, 4$ and $M = c, c + 1, ..., maxM$.

Example 14.3.1

We first consider the solution method of maximizing fuzzy profit. Then we look at ranking the fuzzy numbers for fuzzy profit from smallest to largest.

$max\overline{Profit}$

We will use $\overline{\lambda} = \overline{\lambda}_1 = (3/4/5)$, for the normal fuzzy arrival rate, which was used in Chapters 12 and 13, and for the normal fuzzy server rate, have $\overline{\mu} = \overline{\mu}_1 = (5/6/7)$, also used in Chapters 12 and 13. The faster fuzzy server rate will be $\overline{\mu}_2 = (6/7/8)$ which is the same as in Chapter 13. For the faster fuzzy arrival rate, due to advertising, we choose $\overline{\lambda}_2 = (4/5/6)$.

We will consider the sixteen cases in Table 14.10. These are similar to the 16 cases used in Example 14.2.1. We will number the fuzzy number $\overline{N}_i$, according to the 16 cases $i = 1, 2, 3, ..., 16$ in Table 14.10. Now we go through three steps: (1) find the fuzzy steady state probabilities for Cases 9-16 in Step 1 (we found them for the other cases in Example 13.3.1.1); (2) Step 2 determines α-cuts of $\overline{N}_i$, $1 \leq i \leq 16$; and (3) Step 3 solves the maximization problem of $max\overline{Profit}$.

Case	Number of Servers	System Capacity	Server Rate	Arrival Rate
1	$c = 1$	$M = 4$	$\overline{\mu}_1$	$\overline{\lambda}_1$
2	$c = 1$	$M = 10$	$\overline{\mu}_1$	$\overline{\lambda}_1$
3	$c = 2$	$M = 4$	$\overline{\mu}_1$	$\overline{\lambda}_1$
4	$c = 2$	$M = 10$	$\overline{\mu}_1$	$\overline{\lambda}_1$
5	$c = 1$	$M = 4$	$\overline{\mu}_2$	$\overline{\lambda}_1$
6	$c = 1$	$M = 10$	$\overline{\mu}_2$	$\overline{\lambda}_1$
7	$c = 2$	$M = 4$	$\overline{\mu}_2$	$\overline{\lambda}_1$
8	$c = 2$	$M = 10$	$\overline{\mu}_2$	$\overline{\lambda}_1$
9	$c = 1$	$M = 4$	$\overline{\mu}_1$	$\overline{\lambda}_2$
10	$c = 1$	$M = 10$	$\overline{\mu}_1$	$\overline{\lambda}_2$
11	$c = 2$	$M = 4$	$\overline{\mu}_1$	$\overline{\lambda}_2$
12	$c = 2$	$M = 10$	$\overline{\mu}_1$	$\overline{\lambda}_2$
13	$c = 1$	$M = 4$	$\overline{\mu}_2$	$\overline{\lambda}_2$
14	$c = 1$	$M = 10$	$\overline{\mu}_2$	$\overline{\lambda}_2$
15	$c = 2$	$M = 4$	$\overline{\mu}_2$	$\overline{\lambda}_2$
16	$c = 2$	$M = 10$	$\overline{\mu}_2$	$\overline{\lambda}_2$

Table 14.11: The Sixteen Cases in Example 14.3.1

Step 1

We calculated the alpha-cuts $\overline{w}_i[\alpha]$ of the fuzzy steady state probabilities only for $\alpha = 0, 1/3, 2/3, 1$ to obtain a minimal approximation to the fuzzy numbers $\overline{N}_i$, $1 \leq i \leq 16$. These calculations, for cases 9,11,13,15 were discussed in Chapter 12 and cases 10,12,14,16 were studied in Chapter 13 (the only difference is a different value of $\overline{\lambda}$).

We will not present tables for all these alpha-cuts for all of these fuzzy steady state probabilities. Instead we give only two tables, Table 14.12 for case 9 and Table 14.13 for case 16, for $\alpha = 0$.

$\overline{w}$	Alpha Zero Cut
$\overline{w}_0[0]$	$[0.1344, 0.4564]$
$\overline{w}_1[0]$	$[0.1613, 0.2608]$
$\overline{w}_2[0]$	$[0.1490, 0.2000]$
$\overline{w}_3[0]$	$[0.0852, 0.2322]$
$\overline{w}_4[0]$	$[0.0487, 0.2786]$

Table 14.12: Alpha Zero Cuts of the Fuzzy Steady State Probabilities, Case 9 in Example 14.3.1

$\overline{w}$	Alpha Zero Cut
$\overline{w}_0[0]$	$[0.3335, 0.6000]$
$\overline{w}_1[0]$	$[0.3000, 0.3432]$
$\overline{w}_2[0]$	$[0.0750, 0.1668]$
$\overline{w}_3[0]$	$[0.0188, 0.0834]$
$\overline{w}_4[0]$	$[0.0047, 0.0417]$
$\overline{w}_5[0]$	$[0.0012, 0.0208]$
$\overline{w}_6[0]$	$[0.0003, 0.0104]$
$\overline{w}_7[0]$	$[0.0001, 0.0052]$
$\overline{w}_8[0]$	$[0.0000^*, 0.0026]$
$\overline{w}_9[0]$	$[0.0000^*, 0.0013]$
$\overline{w}_{10}[0]$	$[0.0000^*, 0.0006]$

Table 14.13: Alpha Zero Cuts of the Fuzzy Steady State Probabilities, Case 16 in Example 14.3.1

α	$\overline{N}_9[\alpha]$
1	1.6407
2/3	$[1.4126, 1.8788]$
1/3	$[1.2009, 2.1212]$
0	$[1.0091, 2.3593]$

Table 14.14: Alpha Cuts of $\overline{N}$, Case 9 in Example 14.3.1

Step 2

Here we are to get the alpha-cuts, $\alpha = 0, 1/3, 2/3$, of the $\overline{N}_i$, $9 \leq i \leq 16$. In all cases we find the alpha-cuts of $\overline{N}_i$, by solving a linear programming problem as discussed in Chapter 12. We just change the constraints, equation (12.8) or (12.21), to be

$$w_{i1}(\alpha) \leq w_i \leq w_{i2}(\alpha), 0 \leq i \leq M, w_0 + \ldots + w_M = 1, \qquad (14.9)$$

for $\alpha = 0, 1/3, 2/3$, $M = 4$ or $M = 10$, where $\overline{w}_i[\alpha] = [w_{i1}(\alpha), w_{i2}(\alpha)]$ all i and all α.

Table 14.14 shows the results for $\overline{N}_i$ for Case 9 and Table 14.15 gives the alpha-cuts in Case 16.

Step 3

First we will determine the alpha-cuts, for $\alpha = 0, 1/3, 2/3, 1$, of $\overline{Profit}$ in equation (14.8). We do this using interval arithmetic (see Section 2.3.2 of Chapter 2). This is the same as was done in Example 14.2.1 so we will not

α	$\overline{N}_{16}[\alpha]$
1	0.8180
2/3	[0.7120, 0.9245]
1/3	[0.6139, 1.0582]
0	[0.5337, 1.2371]

Table 14.15: Alpha Cuts of $\overline{N}$, Case 16 in Example 14.3.1

present the details. We use the same fuzzy values of the constants $\overline{A}_u$, $\overline{Q}$, $\overline{C}_v$ and $\overline{T}$ that we employed in Example 14.2.1.

Now $\overline{Profit}$ is a function of c, M, $\overline{\mu}$ and $\overline{\lambda}$ and we show this as writing $\overline{Z} = \overline{Profit}(M, c, \overline{\mu}, \overline{\lambda})$. So, for different values of the variables we get triangular shaped fuzzy numbers $\overline{Z}$. Find the values of the variables to maximize $\overline{Z}$. Let: (1) m_z be the center of the core of $\overline{Z}$; (2) L_z be the area under the graph of the membership function to the left of m_z; and (3) R_z be the area under the graph of the membership function to the right of m_z. See Figure 2.5. In our application the core of $\overline{Z}$ will be a single point. For $max\overline{Profit}$ we substitute: (1) $max[m_z]$; (2) $minL_z$, or minimize the possibility of obtaining values less than m_z; and (3) $maxR_z$, or maximize the possibility of obtaining values greater then m_z. So for $max\overline{Profit}$ we have $max(K_1 - L_z)$, $max[m_z]$ and $max(R_z)$, where K_1 is a sufficiently large positive constant so that $minL_z$ is equivalent to $max(K_1 - L_z)$. We change this multiobjective problem into the single objective

$$max(\lambda_1[K_1 - L_z] + \lambda_2 m_z + \lambda_3 R_z), \tag{14.10}$$

where $\lambda_i > 0$, $1 \leq i \leq 3$, $\lambda_1 + \lambda_2 + \lambda_3 = 1$, and the values of the variables are shown in Table 14.11.

What we need to do is compute L_z and R_z for $\overline{Profit}_i$, $i = 1, 2, ..., 16$. These are needed for the objective function to be maximized. From the α-cuts we have seven points on the graph of the membership function for $\overline{Profit}_i$. We use the Trapezoidal Method, discussed in Example 13.3.1.1, to approximate these areas under the graph of the membership function for the fuzzy number for $\overline{Profit}$. The results are shown in Table 13.16.

Finally, we get to evaluate the objective function to be maximized. In equation (14.10) we will use $K_1 = 2$ since there is a value of L_z in Table 14.16 that exceeds one (but is less than two). Given values of the λ_i, substitute the L_z, m_z and R_z from Table 14.16, we find the value of i (Case number in Table 14.11) that makes the expression in equation (14.10) a maximum. The results for three choices for the λ values is shown in Table 14.17. The (a, b, c) labels at the top of this table means $\lambda_1 = a$, $\lambda_2 = b$ and $\lambda_3 = c$.

The value of the objective function can be misleading because it can be positive when fuzzy profit is approximately negative. We would say that $\overline{Profit}$ is approximately negative when its central value $m_z < 0$. Since $L_z >$

Case	L_z	m_z	R_z
1	0.4251	0.5203	0.5470
2	0.8609	0.6974	1.7374
3	0.2592	−0.2055	0.3327
4	0.3697	−0.5924	0.4849
5	0.3715	0.2587	0.4799
6	0.6272	0.1275	1.1478
7	0.2571	−0.3924	0.3110
8	0.3290	−0.7994	0.4066
9	0.4725	0.7787	0.5601
10	1.3646	1.9243	2.0807
11	0.2893	−0.1327	0.3681
12	0.4227	−0.4739	0.6103
13	0.4219	0.4703	0.5129
14	0.9374	0.8700	1.5068
15	0.2810	−0.3513	0.3405
16	0.3898	−0.7329	0.4514

Table 14.16: Central Value and Certain Areas Under the Graph of $\overline{Profit}_i$, for the Sixteen Cases in Example 14.3.1

Case	$(1/3, 1/3, 1/3)$	$(0.2, 0.4, 0.4)$	$(0.2, 0.5, 0.3)$
1	0.8807	0.7419	0.7392
2	1.1913	1.2018	1.0978
3	0.6227	0.3991	0.3452
4	0.5076	0.2831	0.1754
5	0.7890	0.6212	0.5991
6	0.8827	0.7847	0.6827
7	0.5538	0.3160	0.2457
8	0.4260	0.1771	0.0565
9	0.9554	0.8410	0.8629
10	1.5468	1.7291	1.7134
11	0.6487	0.4363	0.3862
12	0.5712	0.3700	0.2616
13	0.8537	0.7089	0.7046
14	1.1465	1.1633	1.0996
15	0.5693	0.3395	0.2703
16	0.4429	0.2095	0.0910

Table 14.17: The Results of $max\overline{Profit}$ for the Sixteen Cases in Example 14.3.1

0, $R_z > 0$, $K_1 > 0$ and the $\lambda_i > 0$, we may get $\lambda_1[K_1 - L_z] + \lambda_2 m_z + \lambda_3 R_z > 0$ even if $m_z < 0$. We will reject all cases where fuzzy profit is approximately negative. Therefore, from Table 14.16 we omit Cases 3,4,7,8,11,12,15,16, all where there are two servers. Clearly, from those cases left and the results in Table 14.17 Case 10 is best for maximizing fuzzy profit for all values of the λ_i given. Case 10 is $c = 1$, $M = 10$, $\overline{\lambda} = \overline{\lambda}_2$ (the faster fuzzy arrival rate) and $\overline{\mu} = \overline{\mu}_1$ (the slower fuzzy service rate). What happened in Case 10 is that there are more customers expected in the system for these values of the variables producing more profit. More people in the system leads to congestion and larger values of $\overline{R}$. So we might now consider a multi-goal problem of maximizing fuzzy profit together with minimizing $\overline{R}$. These multi-goal fuzzy optimization problems are considered in Chapter 17.

14.3.1   Ranking the Fuzzy Sets

We discussed ranking fuzzy numbers in Section 2.6 of Chapter 2. Let us apply this to $max\overline{\Pi}$. Recall that the highest ranked fuzzy numbers are in the set H_K and the lowest ranked fuzzy numbers are in H_1.

Fuzzy sets $\overline{A}_i$ are the values of $\overline{\Pi}$. Since we wish to maximize $\overline{\Pi}$ we look at all the fuzzy sets in H_K. A problem may arise when we get more than one $\overline{\Pi}$ in H_K. In general we suggest presenting management with all solutions in H_K and let them pick one to implement. If we require a unique solution, then we could go back to the previous section to resolve the conflict.

Example 14.3.1.1

We will only use the $\alpha = 0$ and $\alpha = 1$ cuts of $\overline{\Pi}$ in this section. We do have the $\alpha = 1/3$ and $\alpha = 2/3$ cuts of these fuzzy numbers but using only the $\alpha = 0$ and $\alpha = 1$ cuts will allow us to approximate $\overline{\Pi}$ with triangular fuzzy numbers for ease in calculations and graphing. We did this also in Chapter 13. Using the triangular fuzzy number approximations will still give us a rough idea on how they line up from smallest to largest.

We first round off the $\alpha = 0$ cut and the $\alpha = 1$ cut of $\overline{\Pi}_i$, $1 \leq i \leq 16$, to one decimal place for ease in producing a graph of these fuzzy numbers. Since we only use the $\alpha = 0$ and $\alpha = 1$ cut for these fuzzy numbers we will approximate them by triangular fuzzy numbers. For example, if the alpha equal zero cut of a $\overline{\Pi}$ is $[\pi_1, \pi_3]$ and the alpha equal one cut is π_2, then $\overline{\Pi} \approx (\pi_1/\pi_2/\pi_3)$. We will now drop the $\approx$ and write a $\overline{\Pi}$ as a triangular fuzzy number. These triangular fuzzy numbers for all the $\overline{\Pi}$ are shown in Table 14.18.

Next we graph these triangular fuzzy number from Table 14.18 to see how they line up from smallest to largest. This is now shown in Figure 14.1. In Figure 14.1 we only show the $\overline{\Pi}_i$ in the highest ranked set H_K, defined below. We want H_K the set of largest fuzzy numbers. We will need to

$\overline{\Pi}$	Triangular Fuzzy Number
$\overline{\Pi}_1$	$(-0.3/0.5/1.7)$
$\overline{\Pi}_2$	$(-0.8/0.7/4.6)$
$\overline{\Pi}_3$	$(-0.7/-0.2/0.5)$
$\overline{\Pi}_4$	$(-1.3/-0.6/0.5)$
$\overline{\Pi}_5$	$(-0.4/0.3/1.3)$
$\overline{\Pi}_6$	$(-1.0/0.1/2.8)$
$\overline{\Pi}_7$	$(-0.9/-0.4/0.3)$
$\overline{\Pi}_8$	$(-1.4/-0.8/0.0)$
$\overline{\Pi}_9$	$(-0.1/0.8/1.9)$
$\overline{\Pi}_{10}$	$(-0.5/1.9/6.3)$
$\overline{\Pi}_{11}$	$(-0.7/-0.1/0.6)$
$\overline{\Pi}_{12}$	$(-1.3/-0.5/0.9)$
$\overline{\Pi}_{13}$	$(-0.3/0.5/1.5)$
$\overline{\Pi}_{14}$	$(-0.8/0.9/4.3)$
$\overline{\Pi}_{15}$	$(-0.9/-0.4/0.4)$
$\overline{\Pi}_{16}$	$(-1.5/-0.7/0.2)$

Table 14.18: The Triangular Fuzzy Numbers for the $\overline{\Pi}$ in Example 14.3.1.1

compute $v(\overline{\Pi}_i, \overline{\Pi}_j)$, see Section 2.6 in Chapter 2, for $i \neq j \in \{1, 2, ..., 16\}$. We will use $\eta = 0.8$ in the comparison as discussed in Section 2.6. We found that $H_K = \{\overline{\Pi}_2, \overline{\Pi}_{10}, \overline{\Pi}_{14}\}$. This means that $\overline{\Pi}_2 \approx \overline{\Pi}_{10} \approx \overline{\Pi}_{14}$. Also, if $\overline{\Pi} \in \{\overline{\Pi}_i | i = 1, 3, 4, 5, 6, 7, 8, 9, 11, 12, 13, 15, 16\}$, then $\overline{\Pi} < \overline{\Pi}_{10}$. Hence, H_K is as defined above. Notice also that the height of the intersections between $\overline{\Pi}_2$, $\overline{\Pi}_{10}$ and $\overline{\Pi}_{14}$ is above 0.8 on the vertical axis, so they are all approximately equal to each other. The optimal solution is Cases 2,10 and 14. If we require a unique solution, then we may go back and use the analytical methods in Example 14.4.1. In that example the optimal solution was Case 10.

14.4 References

1. D.A.Menasce and V.A.F.Almeida: Capacity Planning for Web Performance, Prentice Hall, Upper Saddle River, N.J., 1998.

2. Maple 6, Waterloo Maple Inc., Waterloo, Canada.

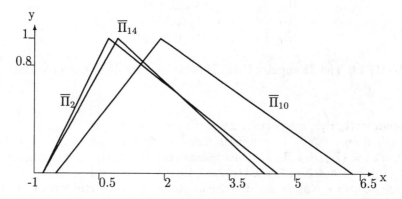

Figure 14.1: Ranking the Fuzzy Numbers $\overline{\Pi}$ in Example 14.3.1.1

Chapter 15

Burstiness

15.1 Introduction

We will model "burstiness" by having two different arrival rates, one for usual operating conditions and a second for when we experience a burst of customers. So we have two cases to consider: (1) where we model arrivals as fuzzy probabilities; and (2) when we use fuzzy arrival and fuzzy service rates.

Burstiness ([3], Chapter 10) is the phenomenon where the system exhibits a large arrival of customers over a short time interval. For example, this might occur just after 1pm and just after 7pm, eastern standard time in the U.S. We will model this situation as a system having two different arrival rates (one for "normal" time and one for "burstiness" time). We first consider the fuzzy queuing system based on fuzzy probabilities.

15.2 Fuzzy Probabilities

We will have two values of $\overline{p}(i)$, the fuzzy probability of i arrivals per unit time, one for normal time $(\overline{p}^{(n)}(i))$ and another for bursty time $(\overline{p}^{(b)}(i))$. These fuzzy probabilities are obtained as discussed in Chapter 3.

We can not design a system just for normal time because then it will become too congested during bursty time and it will take a while to settle back to normal time operations after the burst of customers. Also, we can not design only for bursty time because then the server(s) will be idle too often during normal time. So we will plan a system to be some sort of compromise between the two extremes.

Our analysis in this section will center around $\overline{R}$, average system response time. We could also use $\overline{U}$ or $\overline{X}$. We will use a model that has "ideal" points, which is different from the models employed in Chapters 13 and 14. An ideal point is a " best possible" result of a goal and is usually unattainable. Let us set an ideal goal, or target, for $\overline{R}$ to be $\overline{R}^{(N)}$ for normal time, and a target for $\overline{R}$ during bursty time to be $\overline{R}^{(B)}$.

Let us consider how to set these targets for $\overline{R}$. Both targets will be triangular fuzzy numbers $(a/b/c)$. Determining a, b, c was discussed in Section 3.4 of Chapter 3 for estimating fuzzy parameters using expert opinion. The value of a would be a best response time, b is the acceptable response time and c represents the worst acceptable response time. These numbers are to be obtained for normal (bursty) time giving $\overline{R}^{(N)}$ $(\overline{R}^{(B)})$.

Now we form our compromise ideal goal

$$\overline{R}^{(*)} = \lambda_1 \overline{R}^{(N)} + \lambda_2 \overline{R}^{(B)}, \tag{15.1}$$

for $\lambda_i > 0$ and $\lambda_1 + \lambda_2 = 1$. Suppose $\lambda_1 = 0.65$ and $\lambda_2 = 0.35$, then we place 65% on normal time and 35% on bursty time. We use superscripts for the "ideal" values and subscripts for the fuzzy numbers computed using values of the variables.

The variables are $c = 1, 2, 3$ and $M = 4, ..., 10$. We assume that we have only one type of server so we have only one value of $\overline{p} =$ the fuzzy probability that a customer leaves a server during time period δ, given that the server was busy with this customer at the start of the time period. Given a value of c and M, using $\overline{p}$, let $\overline{R} = \overline{R}_{(N)}$ if we are using $\overline{p}^{(n)}(i)$, and let $\overline{R} = \overline{R}_{(B)}$ when we have $\overline{p}^{(b)}(i)$. Set $\overline{R}_{(*)} = \lambda_1 \overline{R}_{(N)} + \lambda_2 \overline{R}_{(B)}$. We wish to find the values of the variables to get $\overline{R}_{(*)}$ as close as possible to our compromise goal $\overline{R}^{(*)}$. To measure the distance between two fuzzy numbers we need a metric on the set of fuzzy numbers.

A metric D on fuzzy numbers must have the following properties: for fuzzy numbers $\overline{M}, \overline{N}$ and $\overline{P}$

1. $D(\overline{M}, \overline{N}) \geq 0$;

2. $D(\overline{M}, \overline{N}) = D(\overline{N}, \overline{M})$;

3. $D(\overline{M}, \overline{N}) = 0$ if and only if $\overline{M} = \overline{N}$;

4. $D(\overline{M}, \overline{N}) \leq D(\overline{M}, \overline{P}) + D(\overline{P}, \overline{N})$.

The value of D is always a non-negative real number. Metrics for fuzzy numbers are discussed in ([1], Chapter 3). We now need to choose a metric on fuzzy numbers to continue the discussion.

For triangular shaped fuzzy numbers $\overline{M}$ and $\overline{N}$ let $\overline{M}[\alpha] = [m_1(\alpha), m_2(\alpha)]$, $\overline{N}[\alpha] = [n_1(\alpha), n_2(\alpha)]$, $0 \leq \alpha \leq 1$. Next set $L(\alpha) = |m_1(\alpha) - n_1(\alpha)|$ and $R(\alpha) = |m_2(\alpha) - n_2(\alpha)|$. The metric we will use is

$$D(\overline{M}, \overline{N}) = max\{max(L(\alpha), R(\alpha))|0 \leq \alpha \leq 1\}. \tag{15.2}$$

The optimization problem then is

$$minD(\overline{R}_{(*)}, \overline{R}^{(*)}), \tag{15.3}$$

$\overline{p}$	Fuzzy Probability
$\overline{p}(0)$	(0.07/0.10/0.13)
$\overline{p}(1)$	(0.45/0.50/0.55)
$\overline{p}(2)$	(0.07/0.10/0.13)
$\overline{p}(3)$	(0.07/0.10/0.13)
$\overline{p}(4)$	(0.07/0.10/0.13)
$\overline{p}(5)$	(0.07/0.10/0.13)
$\overline{p}(6)$	0
$\overline{p}(7)$	0

Table 15.1: Bursty Fuzzy Probabilities for Arrivals $\overline{p}^{(b)}(i)$ in Example 15.2.1

for $c = 1, 2, 3$ and $M = 4, 5, ..., 10$.

Now we suggest choosing a couple of values for $\overline{R}^{(N)}$, $\overline{R}^{(B)}$ and the λ_i and showing the solutions to management so that they can make the choice on a value for c and M.

If management does not like the "ideal point" approach we can use the methods presented in Chapter 13. The objective would be simply minimize $\overline{R}_{(*)}$.

Example 15.2.1

In this example we will use the fuzzy probabilities in Table 6.1 for our "normal" fuzzy arrival probabilities $\overline{p}^{(n)}(i)$. For the fuzzy server probability we choose $\overline{p} = (0.3/0.4/0.5)$. We used these two $\overline{p}^{(n)}(i)$ and $\overline{p}$ together in previous chapters. For the bursty fuzzy arrival probabilities $\overline{p}^{(b)}(i)$ we choose those given in Table 15.1. Notice that the central value of $\overline{p}^{(b)}(1) = 0.5$ is 1.67 times the central value of $\overline{p}^{(n)}(1) = 0.3$.

For our "ideal" fuzzy targets $\overline{R}^{(N)}$ and $\overline{R}^{(B)}$, both triangular fuzzy numbers, we first pick $\overline{R}^{(N)} = (3.5753/4.7800/6.6365)$. We got the value of $\overline{R}^{(N)}$ from Example 13.2.1.1 in Chapter 13. In this example we wanted to minimize $\overline{R}$ and we found that, see Table 13.9, that for Cases 1-4, which have the same $\overline{p} = (0.3/0.4/0.5)$, Case 3 gave the best results. The $\overline{R}$ for this case became our $\overline{R}^{(N)}$.

Next we need to pick values for the other fuzzy ideal number $\overline{R}^{(B)}$. Not having solved minimize $\overline{R}$ using the new value for the fuzzy arrival probabilities $\overline{p}^{(b)}(i)$ we must guess at two possible values for $\overline{R}^{(B)}$. We first choose $\overline{R}^{(B)} = \overline{R}^{(N)}$. We realize it is not realistic to make $\overline{R}^{(B)}$, depending on the faster fuzzy arrival probabilities, equal to $\overline{R}^{(N)}$, but having no experience with $\overline{p}^{(b)}(i)$ in our optimization models, we feel it is a good place to

Case	Number of Servers	System Capacity	$\overline{p}^{(n)}(i)$	$\overline{p}^{(b)}(i)$
1	$c = 1$	$M = 4$	Table 6.1	Table 15.1
2	$c = 1$	$M = 10$	Table 6.1	Table 15.1
3	$c = 2$	$M = 4$	Table 6.1	Table 15.1
4	$c = 2$	$M = 10$	Table 6.1	Table 15.1

Table 15.2: The Four/Eight Cases in Example 15.2.1

start. For the second choice for $\overline{R}^{(B)}$ we picked triangular fuzzy number (5.9708/7.9826/11.0830) which is approximately 1.67 times $\overline{R}^{(N)}$, since the central value of $\overline{p}^{(b)}(1)$ is 1.67 times the central value of $\overline{p}^{(n)}(1)$.

We will consider the four cases shown in Table 15.2. Notice that in each case we : (1) first use the given value of c and M, $\overline{p} = (0.3/0.4/0.5)$ and $\overline{p}^{(n)}(i)$ to compute α-cuts of $\overline{R}_{(N)}$; and (2) next use the same values of c, M and $\overline{p}$ but employ $\overline{p}^{(b)}(i)$ to obtain α-cuts of $\overline{R}_{(B)}$.

We will number the fuzzy numbers $\overline{U}_i$, $\overline{N}_i$, $\overline{X}_i$ and $\overline{R}_i$ as follows: (1) $i = 1, 2, 3, 4$ for Cases 1,2,3,4 when we use $\overline{p}^{(n)}(i)$; and (2) $i = 5, 6, 7, 8$ for Cases 1,2,3,4 when we have $\overline{p}^{(b)}(i)$. Now we go through four steps: (1) step one is to find the $\alpha = 0, 1$ cuts of the $\overline{p}_{ij}$ in the fuzzy transition matrices $\overline{P}$; (2) Step 2 determines the alpha equal to 0,1 cuts of the fuzzy steady state probabilities; (3) the next step gets the $\alpha = 0, 1$ cuts of $\overline{U}_i,...,\overline{R}_i$, $1 \leq i \leq 8$; and (4) Step 4 solves the minimization problem in equation (15.3) for selected values of the λ_i.

Step 1

We did this using $\overline{p}^{(n)}(i)$ in Chapter 13. We need to now do this again using $\overline{p}^{(b)}(i)$. The calculations for the $\alpha = 0$ cut of the $\overline{p}_{ij}$ are essentially the same for any fuzzy arrival probabilities. The slight change here using $\overline{p}^{(b)}(i)$ are (all items defined in Example 13.2.1.1 in Chapter 13): (1) in Case 2, $c = 1$ and $M = 10$, $h = i = h^* = i^* = 0$; and (2) in Case 4, $c = 2$ and $M = 10$, $I = J = I^* = J^* = 0$. So we now assume we have the $\alpha = 0$ cut for all the $\overline{p}_{ij}$ in all the fuzzy transition matrices. All the fuzzy transition matrices will be fuzzy transition matrices for fuzzy, regular, Markov chains discussed in Section 4.2 of Chapter 4.

Step 2

Here we determine (estimate) the end points of the $\alpha = 0$ cut of the fuzzy steady state probabilities for the cases in Table 15.2. This was accomplished

$\overline{w}$	$\alpha = 0$	$\alpha = 1$
$\overline{w}_0$	[0.0000*,0.0001]	0.0000*
$\overline{w}_1$	[0.0000*,0.0010]	0.0002
$\overline{w}_2$	[0.0007, 0.0098]	0.0029
$\overline{w}_3$	[0.0253, 0.0940]	0.0524
$\overline{w}_4$	[0.8959, 0.0740]	0.9445

Table 15.3: $\alpha = 0, 1$ Cuts of the Fuzzy Steady State Probabilities, Case 1 Using $\overline{p}^{(b)}(i)$ in Example 15.2.1

$\overline{w}$	$\alpha = 0$	$\alpha = 1$
$\overline{w}_0$	[0.0002, 0.0092]	0.0016
$\overline{w}_1$	[0.0021, 0.0584]	0.0137
$\overline{w}_2$	[0.0160, 0.1336]	0.0532
$\overline{w}_3$	[0.0854, 0.2507]	0.1600
$\overline{w}_4$	[0.5794, 0.8941]	0.7716

Table 15.4: $\alpha = 0, 1$ Cuts of the Fuzzy Steady State Probabilities, Case 3 Using $\overline{p}^{(b)}(i)$ in Example 15.2.1

in Chapter 13 using $\overline{p}^{(n)}(i)$. We do it again employing $\overline{p}^{(b)}(i)$. Table 15.3 is for Case 1 and Table 15.4 has the alpha-cuts in Case 3, both using $\overline{p}^{(b)}(i)$.

The $\overline{w}_i[1]$ are easy to get since the $\alpha = 1$ cut of the fuzzy transition matrix is just a crisp matrix P, whose $p_{ij} \in [0, 1]$ and the row sums of P are all one. Therefore P^n converges as $n \to \infty$. The $\overline{w}_i[1]$ are in P^n for sufficiently large n.

Step 3

Now we get the $\alpha = 0, 1$ cuts of $\overline{U}, \overline{N}, \overline{X}$ and $\overline{R}$. The necessary calculations were discussed in detail in previous chapters.

For our optimization model we will need only the $\alpha = 0, 1$ cuts of $\overline{R}$ for all the cases in Table 15.2. The results are presented in Table 15.5.

Step 4

Given a value of λ_1 and λ_2, whose sum is one, we can now compute $\overline{R}^{(*)}$ from equation (15.1) and $\overline{R}_{(*)} = \lambda_1 \overline{R}_{(N)} + \lambda_2 \overline{R}_{(B)}$. Then we may determine $L(\alpha)$, $R(\alpha)$ and $D(\overline{R}^{(*)}, \overline{R}_{(*)})$ from equation (15.2), using only $\alpha = 0, 1$. The results are in Tables 15.6 and 15.7 . The notation (a, b) at the top of these tables means $\lambda_1 = a$ and $\lambda_2 = b$.

$\overline{R}$	$\alpha = 0$	$\alpha = 1$
$\overline{R}_1$	$[7.8074, 13.2507]$	9.869
$\overline{R}_2$	$[19.8072, 33.2507]$	24.869
$\overline{R}_3$	$[3.5753, 6.6365]$	4.7800
$\overline{R}_4$	$[9.5270, 16.5205]$	12.2379
$\overline{R}_5$	$[7.7676, 13.2443]$	9.853
$\overline{R}_6$	$[19.7692, 33.2443]$	24.853
$\overline{R}_7$	$[3.3054, 6.7128]$	4.6474
$\overline{R}_8$	$[9.1557, 16.4502]$	12.0820

Table 15.5: $\alpha = 0, 1$ Cuts of Fuzzy Response Time $\overline{R}$ in Example 15.2.1

Case	$(1/2, 1/2)$	$(0.7, 0.3)$	$(0.3, 0.7)$
1	6.6110	6.6123	6.6097
2	26.6110	26.6123	26.6097
3	0.1350	0.0810	0.1889
4	9.8488	9.8629	9.8348

Table 15.6: The Results of $D(\overline{R}^{(*)}, \overline{R}_{(*)})$, for the Four Cases in Example 15.2.1, using $\overline{R}^{(N)} = \overline{R}^{(B)}$

Case	$(1/2, 1/2)$	$(0.7, 0.3)$	$(0.3, 0.7)$
1	4.3878	5.2783	3.4972
2	24.3878	25.2783	23.4972
3	2.1851	1.3111	3.0591
4	7.6256	8.5290	6.7222

Table 15.7: The Results of $D(\overline{R}^{(*)}, \overline{R}_{(*)})$, for the Four Cases in Example 15.2.1, using $\overline{R}^{(B)} \approx 1.67\overline{R}^{(N)}$

We see from Table 15.6 that when we choose $\overline{R}^{(B)} = \overline{R}^{(N)}$ Case 3 ($c = 2$, $M = 4$) is the best. Recall, we want the values of the variables to minimize the distance (smallest value in each column) between $\overline{R}^{(*)}$ and $\overline{R}_{(*)}$. However, with a larger fuzzy arrival rate $(\overline{\lambda}^{(b)})$ and small system capacity ($M = 4$), we are probably turning many customers away, so we also need to look at the number of lost customers (to be considered in Chapter 17).

Next look at $\overline{R}^{(B)} \approx 1.67\overline{R}^{(N)}$ whose results are in Table 15.7. For all choices of the λ_i given, Case 3 was again best with $c = 2$ and $M = 10$.

This is probably the better result of the two choices of $\overline{R}^{(B)}$ since it is more realistic to pick the "ideal" fuzzy set for burstiness $\overline{R}^{(B)}$ larger than the "ideal" fuzzy set for normal operations $\overline{R}^{(N)}$.

15.2.1 Ranking the Fuzzy Numbers

Here we assume we wish to minimize $\overline{R}_{(*)}$. So let $\overline{R}_{(*)}^i = \lambda_1\overline{R}_{(N)} + \lambda_2\overline{R}_{(B)}$ for $i = 1, 2, 3, 4$ where the values of $\overline{R}_{(N)}$ and $\overline{R}_{(B)}$ are determined as in Case i in Table 15.2. For example, if $i = 3$ we use the values for $\overline{R}_{(N)}$ and $\overline{R}_{(B)}$ computed in Case 3 in the above equation to obtain $\overline{R}_{(*)}^3$. Given values for the λ_i we then rank the fuzzy numbers $\overline{R}_{(*)}^i$ from smallest to largest and pick out those in the smallest set H_1 for the optimal solution.

Example 15.2.1.1

In this example we use $\lambda_1 = \lambda_2 = 0.5$. We will only use the $\alpha = 0$ and $\alpha = 1$ cuts of $\overline{R}_{(*)}^i$ in this section. We do have the $\alpha = 1/3$ and $\alpha = 2/3$ cuts of these fuzzy numbers but using only the $\alpha = 0$ and $\alpha = 1$ cuts will allow us to approximate $\overline{R}_{(*)}^i$ with triangular fuzzy numbers for ease in calculations and graphing. We did this in previous chapters. Using the triangular fuzzy number approximations will still give us a rough idea on how they line up from smallest to largest.

We first round off the $\alpha = 0$ cut and the $\alpha = 1$ cut of $\overline{R}_i$, $1 \leq i \leq 8$, in Table 15.5 to two decimal places. Then compute $\overline{R}_{(*)}^j$, $1 \leq j \leq 4$. We will now write a $\overline{R}_{(*)}^j$ as a triangular fuzzy number $(r_1/r_2/r_3)$ where $[r_1, r_3]$ is the alpha zero cut and r_2 is the alpha one cut. The triangular fuzzy numbers for the $\overline{R}_{(*)}^j$ are in Table 15.8. In Table 15.8 Case j corresponds to $\overline{R}_{(*)}^j$, $j = 1, 2, 3, 4$.

Next we graph these triangular fuzzy numbers from Table 15.8 to see how they line up from smallest to largest. This is now shown in Figure 15.1. In Figure 15.1 we obtain $H_1 = \{Case\ \ 3\}$, and the optimal solution is Cases 3 when $\lambda_1 = \lambda_2 = 0.5$. This is the same result obtained in Example

$\overline{R}_{(*)}^i$	Triangular Fuzzy Number
Case 1	(7.79/9.86/13.24)
Case 2	(19.79/24.86/33.24)
Case 3	(3.44/4.72/6.68)
Case 4	(9.34/12.16/16.48)

Table 15.8: The Triangular Fuzzy Numbers for the $\overline{R}_{(*)}^i$ in Example 15.2.1.1

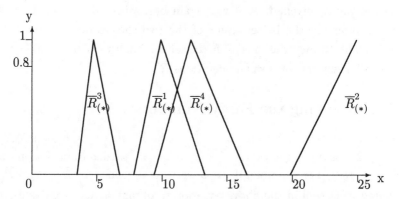

Figure 15.1: Ranking the Fuzzy Numbers $\overline{R}_{(*)}^i$ in Example 15.2.1.1

15.2.1. Using the horizontal line through 0.8 criterion we easily see that $\overline{R}_{(*)}^3 < \overline{R}_{(*)}^1 < \overline{R}_{(*)}^4 < \overline{R}_{(*)}^2$.

15.3   Fuzzy Arrival/Service Rates

We will have two values of $\overline{\lambda}$, the fuzzy arrival rate, one for normal time $(\overline{\lambda}^{(n)})$ and another for bursty time $(\overline{\lambda}^{(b)})$. These fuzzy arrival rates are obtained as discussed in Chapter 3.

Our analysis will center around $\overline{R}$, average system response time. We will use a model that has "ideal" points. An ideal point is a " best possible" result of a goal and is usually unattainable. Let us set an ideal goal, or target, for $\overline{R}$ to be $\overline{R}^{(N)}$ for normal time, and a target for $\overline{R}$ during bursty time to be $\overline{R}^{(B)}$.

Now we form our compromise ideal goal

$$\overline{R}^{(*)} = \lambda_1 \overline{R}^{(N)} + \lambda_2 \overline{R}^{(B)}, \tag{15.4}$$

for $\lambda_i > 0$ and $\lambda_1 + \lambda_2 = 1$. Suppose $\lambda_1 = 0.65$ and $\lambda_2 = 0.35$, then we place 65% on normal time and 35% on bursty time. We use superscripts for the "ideal" values and subscripts for the fuzzy numbers computed using values of the variables.

The variables are $c = 1, 2, 3$ and $M = 4, ..., 10$. We assume that we have only one type of server so we have only one value of $\overline{\mu}$ = the fuzzy service rate. Given a value of c and M, using $\overline{\mu}$, let $\overline{R} = \overline{R}_{(N)}$ if we are using $\overline{\lambda}^{(n)}$, and let $\overline{R} = \overline{R}_{(B)}$ when we have $\overline{\lambda}^{(b)}$. Set $\overline{R}_{(*)} = \lambda_1 \overline{R}_{(N)} + \lambda_2 \overline{R}_{(B)}$. We wish to find the values of the variables to get $\overline{R}_{(*)}$ as close as possible to our compromise goal $\overline{R}^{(*)}$. To measure the distance between two fuzzy numbers we use a metric D on the set of fuzzy numbers.

The optimization problem then is

$$minD(\overline{R}_{(*)}, \overline{R}^{(*)}),\qquad(15.5)$$

for $c = 1, 2, 3$ and $M = 4, ..., 10$.

Example 15.3.1

We will use, for the normal fuzzy arrival rate, $\overline{\lambda}^{(n)} = (3/4/5)$ and $\overline{\mu} = (5/6/7)$ for the fuzzy service rate. Both values were used in previous chapters. For the faster fuzzy arrival rate due to "burstiness" we choose $\overline{\lambda}^{(b)} = (6/7/8)$ which is approximately 1.75 times the normal fuzzy arrival rate.

For our "ideal" fuzzy targets $\overline{R}^{(N)}$ and $\overline{R}^{(B)}$, both triangular fuzzy numbers, we first pick $\overline{R}^{(N)} = (0.0665/0.1824/0.5291)$. We got the value of $\overline{R}^{(N)}$ from Example 13.3.1.1 in Chapter 13. In this example we wanted to minimize $\overline{R}$ and we found that, see Table 13.20, that for $\overline{\mu} = \overline{\mu}_1 = (5/6/7)$, Case 3 ($c = 2$, $M = 4$) was the best. For these values of $c = 2$, $M = 4$, fuzzy arrival rate of $(3/4/5)$ and fuzzy service rate equal $(5/6/7)$ we had $\overline{R}[0] = [0.0665, 0.5291]$ and $\overline{R}[1] = 0.1824$. These were used to produce the fuzzy triangular number for $\overline{R}^{(N)}$.

Next we need to pick values for the other fuzzy ideal number $\overline{R}^{(B)}$. Not having solved minimize $\overline{R}$ using the new value for the fuzzy arrival rate $\overline{\lambda}^{(b)}$ we must guess at two possible values for $\overline{R}^{(B)}$. We first choose $\overline{R}^{(B)} = \overline{R}^{(N)}$. We realize it is not realistic to make $\overline{R}^{(B)}$, depending on the faster fuzzy arrival rate, equal to $\overline{R}^{(N)}$, but having no experience with $\overline{\lambda}^{(b)}$ in our optimization models, we feel it is a good place to start. For the second choice for $\overline{R}^{(B)}$ we picked triangular fuzzy number $(0.1164/0.3192/0.9259)$ which is approximately 1.75 times $\overline{R}^{(N)}$, since $\overline{\lambda}^{(b)} \approx 1.75\overline{\lambda}^{(n)}$.

We will consider the four cases shown in Table 15.9. These are similar to the cases used in Example 15.2.1. Notice that in each case we : (1) first use the given value of c and M, $\overline{\mu} = (5/6/7)$ and $\overline{\lambda}^{(n)}$ to compute α-cuts of $\overline{R}_{(N)}$;

Case	Number of Servers	System Capacity	$\overline{\lambda}^{(n)}$	$\overline{\lambda}^{(b)}$
1	$c = 1$	$M = 4$	(3/4/5)	(6/7/8)
2	$c = 1$	$M = 10$	(3/4/5)	(6/7/8)
3	$c = 2$	$M = 4$	(3/4/5)	(6/7/8)
4	$c = 2$	$M = 10$	(3/4/5)	(6/7/8)

Table 15.9: The Four Cases in Example 15.3.1

and (2) next use the same values of c, M and $\overline{\mu}$ but employ $\overline{\lambda}^{(b)}$ to obtain α-cuts of $\overline{R}_{(B)}$.

We will number the fuzzy numbers $\overline{U}_i$, $\overline{N}_i$, $\overline{X}_i$ and $\overline{R}_i$ as follows: (1) $i = 1, 2, 3, 4$ for Cases 1,2,3,4 when we use $\overline{\lambda}^{(n)}$; and (2) $i = 5, 6, 7, 8$ for Cases 1,2,3,4 when we have $\overline{\lambda}^{(b)}$. Now we go through three steps: (1) find the needed fuzzy steady state probabilities in Step 1; (2) Step 2 determines the $\overline{U}_i, ..., \overline{R}_i$, $5 \leq i \leq 8$; and (3) Step 3 solves the minimization problem in equation (15.5) for selected values of the λ_i.

Step 1

We calculated the alpha-cuts $\overline{w}_i[\alpha]$ of the fuzzy steady state probabilities, for $\alpha = 0, 1/3, 2/3, 1$, for Cases 1,2,3,4 using $\overline{\lambda}^{(n)}$, in Example 13.3.1.1. Therefore, we only need to do this using $\overline{\lambda}^{(b)}$. We did this using the Premium Solver Platform V5.0 from Frontline Systems [2] as explained after Example 12.2.1 in Chapter 12. Table 15.10 presents the $\alpha = 0$ cut for Case 2 and $\overline{\lambda}^{(b)}$ while Table 15.11 is for the alpha zero cut in Case 3 for $\overline{\lambda}^{(b)}$.

Step 2

Here we are to get the alpha-cuts, $\alpha = 0, 1/3, 2/3$, of the $\overline{U}_i$, $\overline{N}_i$, $\overline{X}_i$ and $\overline{R}_i$, $1 \leq i \leq 8$. We do not need the $\overline{U}_i$ in this example. We obtained these alpha-cuts using $\overline{\lambda}^{(n)}$ in Example 13.3.1.1 for $i = 1, 2, 3, 4$. So we need to do this for $\overline{\lambda}^{(b)}$ for $i = 5, 6, 7, 8$. In all cases we find the alpha-cuts of $\overline{U}_i$, $\overline{N}_i$ and $\overline{X}_i$ by solving a linear programming problem as discussed in Chapter 12. We just change the constraints, equation (12.8) or (12.21), to be

$$w_{i1}(\alpha) \leq w_i \leq w_{i2}(\alpha), 0 \leq i \leq M, w_0 + ... + w_M = 1, \qquad (15.6)$$

for $\alpha = 0, 1/3, 2/3$, $M = 4$ or $M = 10$, where $\overline{w}_i[\alpha] = [w_{i1}(\alpha), w_{i2}(\alpha)]$ all i and all α. $\overline{R}$ is $\overline{N}/\overline{X}$.

There is another change in computing the $\overline{X}_i$. Let $\overline{\mu}[\alpha] = [\mu_1(\alpha), \mu_2(\alpha)]$ for $\alpha = 0, 1/3, 2/3$. If $c = 1$, then: (1) in equation (12.11) use $\mu_1(\alpha)$ for the

$\overline{w}$	Alpha Zero Cut
$\overline{w}_0[0]$	$[0.0034, 0.1749]$
$\overline{w}_1[0]$	$[0.0055, 0.1500]$
$\overline{w}_2[0]$	$[0.0088, 0.1285]$
$\overline{w}_3[0]$	$[0.0140, 0.1102]$
$\overline{w}_4[0]$	$[0.0225, 0.0956]$
$\overline{w}_5[0]$	$[0.0360, 0.0909]$
$\overline{w}_6[0]$	$[0.0575, 0.0956]$
$\overline{w}_7[0]$	$[0.0595, 0.1120]$
$\overline{w}_8[0]$	$[0.0510, 0.1501]$
$\overline{w}_9[0]$	$[0.0437, 0.2357]$
$\overline{w}_{10}[0]$	$[0.0374, 0.3771]$

Table 15.10: Alpha Zero Cuts of the Fuzzy Steady State Probabilities, Case 2 and $\overline{\lambda}^{(b)}$, in Example 15.3.1

$\overline{w}$	Alpha Zero Cut
$\overline{w}_0[0]$	$[0.1747, 0.4083]$
$\overline{w}_1[0]$	$[0.2796, 0.3503]$
$\overline{w}_2[0]$	$[0.1500, 0.2236]$
$\overline{w}_3[0]$	$[0.0643, 0.1789]$
$\overline{w}_4[0]$	$[0.0275, 0.1431]$

Table 15.11: Alpha Zero Cuts of the Fuzzy Steady State Probabilities, Case 3 and $\overline{\lambda}^{(b)}$, Example 15.3.1

α	$\overline{N}[\alpha]$	$\overline{X}[\alpha]$	$\overline{R}[\alpha]$
1	6.4721	5.7756	1.1206
2/3	$[5.5106, 7.2857]$	$[5.2734, 6.2174]$	$[0.8863, 1.3816]$
1/3	$[4.4884, 7.9220]$	$[4.7146, 6.6120]$	$[0.6788, 1.6803]$
0	$[3.5161, 8.4485]$	$[4.1255, 6.9762]$	$[0.5040, 2.0479]$

Table 15.12: Alpha Cuts of $\overline{N}$, $\overline{X}$ and $\overline{R}$, Case 2 and $\overline{\lambda}^{(b)}$, in Example 15.3.1

α	$\overline{N}[\alpha]$	$\overline{X}[\alpha]$	$\overline{R}[\alpha]$
1	1.3362	4.2738	0.3126
2/3	$[1.1959, 1.4904]$	$[3.8097, 4.7595]$	$[0.2513, 0.3912]$
1/3	$[1.0683, 1.5598]$	$[3.3706, 5.1187]$	$[0.2087, 0.4628]$
0	$[0.9528, 1.8360]$	$[2.9585, 5.7771]$	$[0.1649, 0.6206]$

Table 15.13: Alpha Cuts of $\overline{N}$, $\overline{X}$ and $\overline{R}$, Case 3 and $\overline{\lambda}^{(b)}$, in Example 15.3.1

"5"; and (2) in equation (12.12) use $\mu_2(\alpha)$ for the "7". If $c = 2$, then: (1) in equation (12.24) use $\mu_1(\alpha)$ for the "5" and two times this μ value for the "10"; and (2) in equation (12.25) use $\mu_2(\alpha)$ for the "7" and twice this μ value for the "14". Of course, we need to do this when alpha is $0, 1/3, 2/3$. Table 15.11 shows the results for $\overline{N}$, $\overline{X}$ and $\overline{R}$ for Case 2 for $\overline{\lambda}^{(b)}$ and Table 15.12 gives the alpha-cuts of these fuzzy numbers in Case 3 using $\overline{\lambda}^{(b)}$.

Step 3

Given a value of λ_1 and λ_2, whose sum is one, we can now compute $\overline{R}^{(*)}$ from equation (15.4) and $\overline{R}_{(*)} = \lambda_1 \overline{R}_{(N)} + \lambda_2 \overline{R}_{(B)}$. Then we may determine $L(\alpha)$, $R(\alpha)$ and $D(\overline{R}^{(*)}, \overline{R}_{(*)})$ from equation (15.5), using only $\alpha = 0, 1/3, 2/3, 1$. The results are in Tables 15.14 and 15.15 . The notation (a, b) at the top of these tables means $\lambda_1 = a$ and $\lambda_2 = b$.

We see from Table 15.14 that when we choose $\overline{R}^{(B)} = \overline{R}^{(N)}$ Case 3 ($c = 2$, $M = 4$) is the best. Recall, we want the values of the variables to minimize the distance (smallest value in each column) between $\overline{R}^{(*)}$ and $\overline{R}_{(*)}$. However, with a larger fuzzy arrival rate $(\overline{\lambda}^{(b)})$ and small system capacity ($M = 4$), we are probably turning many customers away, so we also need to look at the number of lost customers (to be considered in Chapter 17).

Next look at $\overline{R}^{(B)} \approx 1.75 \overline{R}^{(N)}$ whose results are in Table 15.15. For all choices of the λ_i given, Case 4 was best with $c = 2$ and $M = 10$. This is probably the better result of the two choices of $\overline{R}^{(B)}$ since it is more realistic

Case	(1/2, 1/2)	(0.7, 0.3)	(0.3, 0.7)
1	0.3364	0.3710	0.3019
2	1.6628	1.7203	1.6052
3	0.0651	0.0391	0.0911
4	0.0911	0.1374	0.2008

Table 15.14: The Results of $D(\overline{R}^{(*)}, \overline{R}_{(*)})$, for the Four Cases in Example 15.3.1, using $\overline{R}^{(N)} = \overline{R}^{(B)}$

Case	(1/2, 1/2)	(0.7, 0.3)	(0.3, 0.7)
1	0.1381	0.2520	0.1158
2	1.4644	1.6013	1.3274
3	0.1536	0.1107	0.2137
4	0.1056	0.0714	0.1445

Table 15.15: The Results of $D(\overline{R}^{(*)}, \overline{R}_{(*)})$, for the Four Cases in Example 15.3.1, using $\overline{R}^{(B)} \approx 1.75\overline{R}^{(N)}$

to pick the "ideal" fuzzy set for burstiness $\overline{R}^{(B)}$ larger than the "ideal" fuzzy set for normal operations $\overline{R}^{(N)}$.

15.3.1 Ranking the Fuzzy Numbers

Here we assume we wish to minimize $\overline{R}_{(*)}$. So let $\overline{R}_{(*)}^i = \lambda_1 \overline{R}_{(N)} + \lambda_2 \overline{R}_{(B)}$ for $i = 1, 2, 3, 4$ where the values of $\overline{R}_{(N)}$ and $\overline{R}_{(B)}$ are determined as in Case i in Table 15.9. For example, if $i = 3$ we use the values for $\overline{R}_{(N)}$ and $\overline{R}_{(B)}$ computed in Case 3 in the above equation to obtain $\overline{R}_{(*)}^3$. Given values for the λ_i we then rank the fuzzy numbers $\overline{R}_{(*)}^i$ from smallest to largest and pick out those in the smallest set H_1 for the optimal solution.

Example 15.3.1.1

In this example we use $\lambda_1 = \lambda_2 = 0.5$. We will only use the $\alpha = 0$ and $\alpha = 1$ cuts of $\overline{R}_{(*)}^i$ in this section. We do have the $\alpha = 1/3$ and $\alpha = 2/3$ cuts of these fuzzy numbers but using only the $\alpha = 0$ and $\alpha = 1$ cuts will allow us to approximate $\overline{R}_{(*)}^i$ with triangular fuzzy numbers for ease in calculations and graphing. We did this in previous chapters. Using the triangular fuzzy number approximations will still give us a rough idea on how they line up from smallest to largest.

$\overline{R}^i_{(*)}$	Triangular Fuzzy Number
Case 1	(0.19/0.38/0.86)
Case 2	(0.30/0.80/2.20)
Case 3	(0.12/0.24/0.58)
Case 4	(0.08/0.22/0.70)

Table 15.16: The Triangular Fuzzy Numbers for the $\overline{R}^i_{(*)}$ in Example 15.3.1.1

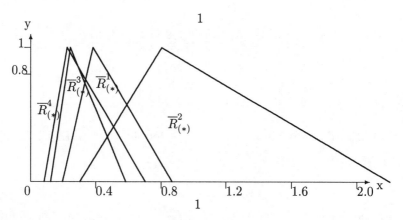

Figure 15.2: Ranking the Fuzzy Numbers $\overline{R}^i_{(*)}$ in Example 15.3.1.1

We first round off the $\alpha = 0$ cut and the $\alpha = 1$ cut of $\overline{R}^i_{(*)}$, $1 \leq i \leq 4$, to two decimal places for ease in producing a graph of these fuzzy numbers. We will now write a $\overline{R}^i_{(*)}$ as a triangular fuzzy number $(r_1/r_2/r_3)$ where $[r_1, r_3]$ is the alpha zero cut and r_2 is the alpha one cut. The triangular fuzzy numbers for the $\overline{R}^i_{(*)}$ are in Table 15.16.

Next we graph these triangular fuzzy numbers from Table 15.16 to see how they line up from smallest to largest. This is now shown in Figure 15.2. In Figure 15.2 we obtain $H_1 = \{Case\,3, Case\,4\}$, $H_2 = \{Case\,1\}$ and $H_3 = \{Case\,2\}$. The optimal solution is Cases 3 and 4 when $\lambda_1 = \lambda_2 = 0.5$ which were optimal in Example 15.3.1. Looking at the heights of the intersections we clearly see in Figure 15.2 that $\overline{R}^4_{(*)} \approx \overline{R}^3_{(*)} < \overline{R}^1_{(*)} < \overline{R}^2_{(*)}$.

15.4   References

1. J.J.Buckley and E.Eslami: Introduction to Fuzzy Logic and Fuzzy Sets, Physica-Verlag, Heidelberg, Germany, 2002.

2. Frontline Systems (www.frontsys.com).

3. D.A.Menasce and V.A.F.Almeida: Capacity Planning for Web Performance, Prentice Hall, Upper Saddle River, N.J., 1998.

Chapter 16

Long Tailed Distributions

16.1 Introduction

By "long tailed distributions" ([2], Chapter 10) we mean that we have different classes of customers, with one class of customer taking a long time in a server. We will consider only two classes of customer: (1) normal, or those who take a regular, or normal, time in a server; and (2) long, or those who usually take a much longer time in a server. The long tailed distribution relates to the fact that there is a positive probability that a customer will take a long time in the server. We will model this situation as having two different arrival rates, one for the "normal" customers and another for those who take a long time in the servers. Also, there will be two different service rates. The results will depend on whether we are using fuzzy probabilities or fuzzy arrival/service rates.

16.2 Fuzzy Probabilities

We will have two values of $\overline{p}(i)$: (1) $\overline{p}^{(n)}(i)$ for the normal ; and (2) $\overline{p}^{(l)}(i)$ for the "long" customers. We will also have two values for $\overline{p}$: (1) $\overline{p}^{(n)}$ for our normal customers; and (2) $\overline{p}^{(l)}$ for those who take much longer in a server. We could use the "ideal point" method presented in the previous chapter, but instead we will employ the optimization techniques from Chapters 13 and 14. We also only concentrate on minimizing $\overline{R}$.

For a value of c and M let $\overline{R} = \overline{R}^{(N)}$ when we use $\overline{p}^{(n)}(i)$ and $\overline{p}^{(n)}$, and for the same values of c and M let $\overline{R} = \overline{R}^{(L)}$ when we have $\overline{p}^{(l)}(i)$ and $\overline{p}^{(l)}$. We form our compromise goal $\overline{Z} = \mu_1 \overline{R}^{(n)} + \mu_2 \overline{R}^{(L)}$. Our objective would be to $min\overline{Z}$ for $c = 1, 2, 3$ and $M = 4, 5, ..., 10$.

$\overline{p}$	Fuzzy Probability
$\overline{p}(0)$	(0.40/0.50/0.60)
$\overline{p}(1)$	(0.40/0.50/0.60)
$\overline{p}(2)$	0
$\overline{p}(3)$	0
$\overline{p}(4)$	0
$\overline{p}(5)$	0
$\overline{p}(6)$	0
$\overline{p}(7)$	0

Table 16.1: Long Tailed Fuzzy Arrival Probabilities $\overline{p}^{(l)}(i)$ in Example 16.2.1

$\overline{q}$	$\alpha = 0$	$\alpha = 1$
$\overline{q}(0|2)[\alpha]$	[0.49, 0.81]	0.64
$\overline{q}(1|2)[\alpha]$	[0.18, 0.42]	0.32
$\overline{q}(2|2)[\alpha]$	[0.01, 0.09]	0.04

Table 16.2: $\alpha = 0, 1$ Cuts of $\overline{q}(i|2)$ Using $\overline{p}^{(l)}(i) = (0.1/0.2/0.3)$ in Example 16.2.1

Example 16.2.1

For the "normal" fuzzy arrival probabilities $\overline{p}^{(n)}(i)$ we will use those in Table 6.1 and $\overline{p}^{(n)} = (0.3/0.4/0.5)$ will be the "normal" fuzzy service probability. We used these two fuzzy numbers together in previous chapters. With respect to the long tailed distribution the fuzzy arrival probabilities $\overline{p}^{(l)}(i)$ are shown in Table 16.1 and the corresponding fuzzy service probability $\overline{p}^{(l)} = (0.1/0.2/0.3)$. Notice that $\overline{p}^{(l)} \approx 0.5\overline{p}^{(n)}$. We need the $\alpha = 0$ cuts of $\overline{q}(i|2)$ using this new value of $\overline{p}$ and they are given in Table 16.2.

We will consider four cases shown in Table 16.3. Notice that in each case we : (1) first use the given value of c and M, $\overline{p}^{(n)}(i)$ from Table 6.1 and $\overline{p}^{(n)} = (0.3/0.4/0.5)$ to compute $\alpha = 0, 1$ cuts of $\overline{R}^{(N)}$; and (2) next use the same values of c, M, $\overline{p}^{(l)}(i)$ from Table 16.1 and $\overline{p}^{(l)} = (0.1/0.2/0.3)$ to obtain $\alpha = 0, 1$ cuts of $\overline{R}^{(L)}$.

We will number the fuzzy numbers $\overline{U}_i$, $\overline{N}_i$, $\overline{X}_i$ and $\overline{R}_i$ as follows: (1) $i = 1, 3, 5, 7$ for Cases 1,2,3,4 when we use $\overline{p}^{(n)}(i)$ and $\overline{p}^{(n)}$; and (2) $i = 2, 4, 6, 8$ for Cases 1,2,3,4 when we have $\overline{p}^{(l)}(i)$ and $\overline{p}^{(l)}$. Now we go through five steps: (1) calculate the $\alpha = 0, 1$ cuts of the $\overline{p}_{ij}$ in the fuzzy transition matrices $\overline{P}$; (2) find the $\alpha = 0, 1$ cuts of the fuzzy steady state probabilities ; (3) determine $\overline{U}_i, ..., \overline{R}_i$, $1 \le i \le 8$; (4) compute $\overline{Z} = \mu_1 \overline{R}^{(N)} + \mu_2 \overline{R}^{(L)}$, for selected values of the μ_i and then find the central value of $\overline{Z}$, called m_z, and compute L_z (R_z) the area under the graph of the membership function to the left (right)

Case	Number of Servers	System Capacity	$\overline{p}(i)$	$\overline{p}$	$\overline{R}$
1	$c = 1$	$M = 4$	$\overline{p}^{(n)}(i)$	$\overline{p}^{(n)}$	$\overline{R}^{(N)}$
			$\overline{p}^{(l)}(i)$	$\overline{p}^{(l)}$	$\overline{R}^{(L)}$
2	$c = 1$	$M = 10$	$\overline{p}^{(n)}(i)$	$\overline{p}^{(n)}$	$\overline{R}^{(N)}$
			$\overline{p}^{(l)}(i)$	$\overline{p}^{(l)}$	$\overline{R}^{(L)}$
3	$c = 2$	$M = 4$	$\overline{p}^{(n)}(i)$	$\overline{p}^{(n)}$	$\overline{R}^{(N)}$
			$\overline{p}^{(l)}(i)$	$\overline{p}^{(l)}$	$\overline{R}^{(L)}$
4	$c = 2$	$M = 10$	$\overline{p}^{(n)}(i)$	$\overline{p}^{(n)}$	$\overline{R}^{(N)}$
			$\overline{p}^{(l)}(i)$	$\overline{p}^{(l)}$	$\overline{R}^{(L)}$

Table 16.3: The Four/Eight Cases in Example 16.2.1

of m_z; and (5) solve the minimization problem

$$min\{\lambda_1(K_1 - L_z) + \lambda_2 m_z + \lambda_3 R_z\}, \tag{16.1}$$

the same as equation (13.3), for selected values of the λ_i.

Step 1

We found the $\overline{p}_{ij}[\alpha]$ for $\alpha = 0, 1$ in $\overline{P}$ for $\overline{p}^{(n)}(i)$ and $\overline{p}^{(n)}$ in previous examples. Now we need to do this for $\overline{p}^{(l)}(i)$ and $\overline{p}^{(l)}$.

Since the crisp $p^{(l)}(i) = 0$ for $i \geq 2$ the crisp transition matrices will contain many zeros. The results are: (1) for $c = 1$ and $M = 4$, $p(2) = p(3) = p(4) = p^*(2) = p^*(3) = p^*(4) = 0$ in Table 6.2; (2) for $c = 1$ and $M = 10$, $p(2) = p(3) = p(4) = p(5) = p(6) = p(7) = p^*(2) = p^*(3) = p^*(4) = p^*(5) = p^*(6) = 0$ in the definitions of $a, b, c, ..., b^*$ in Example 13.2.1.1 in Chapter 13; (3) for $c = 2$ and $M = 4$, $p(2) = p(3) = p(4) = p^*(2) = p^*(3) = p^*(4) = 0$ in Table 8.1; and (4) for $c = 2$ and $M = 10$, $p(2) = p(3) = p(4) = p(5) = p(6) = p(7) = p^*(2) = p^*(3) = p^*(4) = p^*(5) = p^*(6) = 0$ in the definitions of $A, B, ..., C^*$ in Example 13.2.1.1 in Chapter 13. The resulting non-zero $\overline{p}_{ij}[0]$ were computed as discussed in previous examples.

Step 2

The $\alpha = 0, 1$ cuts of the fuzzy steady state probabilities have already been computed under "normal" operating conditions. Therefore we now determine the alpha equal to zero and one cuts of these fuzzy probabilities for the "long tail distribution" parameters. Table 16.4 gives the $\alpha = 0, 1$ cuts for Case 4 using $\overline{p}^{(l)}(i)$ and $\overline{p}^{(l)}$ while Table 16.5 presents the $\alpha = 0, 1$ cuts in Case 1 using $\overline{p}^{(l)}(i)$ and $\overline{p}^{(l)}$.

$\overline{w}$	Alpha Zero Cut	Alpha One Cut
$\overline{w}_0$	[0.0000*,0.1251]	0.0029
$\overline{w}_1$	[0.0000*,0.2622]	0.0109
$\overline{w}_2$	[0.0001,0.3031]	0.0191
$\overline{w}_3$	[0.0003,0.2963]	0.0269
$\overline{w}_4$	[0.0009,0.2768]	0.0377
$\overline{w}_5$	[0.0027,0.2816]	0.0529
$\overline{w}_6$	[0.0075,0.2834]	0.0742
$\overline{w}_7$	[0.0154,0.3075]	0.1041
$\overline{w}_8$	[0.0111,0.3290]	0.1464
$\overline{w}_9$	[0.0065,0.3737]	0.2019
$\overline{w}_{10}$	[0.0045,0.6647]	0.3230

Table 16.4: Alpha Cuts of the Fuzzy Steady State Probabilities, Case 4, $\overline{p}^{(l)}(i)$, $\overline{p}^{(l)}$, in Example 16.2.1

$\overline{w}$	Alpha Zero Cut	Alpha one Cut
$\overline{w}_0$	[0.0000*,0.0490]	0.0023
$\overline{w}_1$	[0.0004,0.1114]	0.0117
$\overline{w}_2$	[0.0051,0.1904]	0.0469
$\overline{w}_3$	[0.0677,0.3527]	0.1878
$\overline{w}_4$	[0.4096,0.9259]	0.7512

Table 16.5: Alpha Cuts of the Fuzzy Steady State Probabilities, Case 1, $\overline{p}^{(l)}(i)$, $\overline{p}^{(l)}$, in Example 16.2.1

All the fuzzy transition matrices will be fuzzy transition matrices for fuzzy, regular, Markov chains discussed in Section 4.2 of Chapter 4.

Step 3

Here we are to get the alpha-cuts, $\alpha = 0, 1$, of the $\overline{U}_i$, $\overline{N}_i$, $\overline{X}_i$ and $\overline{R}_i$, $1 \leq i \leq 8$. We obtained these alpha-cuts using $\overline{p}^{(n)}(i)$ and $\overline{p}^{(n)}$ in previous examples, for $i = 1, 3, 5, 7$. So we need to do this for $\overline{p}^{(l)}(i)$ and $\overline{p}^{(l)}$ for $i = 2, 4, 6, 8$. In all cases we find the alpha-cuts of $\overline{U}_i$, $\overline{N}_i$ and $\overline{X}_i$ by solving a linear programming problem as discussed in previous chapters, so we do not present these details again. $\overline{R}$ is just $\overline{N}/\overline{X}$ evaluated using interval arithmetic.

Table 16.6 shows the results for $\overline{N}$, $\overline{X}$ and $\overline{R}$ for Case 4 for $\overline{p}^{(l)}(i)$ and $\overline{p}^{(l)}$ and Table 13.7 gives the alpha-cuts of these fuzzy numbers in Case 1 when we are using $\overline{p}^{(l)}(i)$ and $\overline{p}^{(l)}$.

α	$\overline{N}_8[\alpha]$	$\overline{X}_8[\alpha]$	$\overline{R}_8[\alpha]$
1	7.9373	0.3967	20.0103
0	[2.0136, 9.5825]	[0.1488, 0.6000]	[3.3560, 64.3985]

Table 16.6: Alpha Cuts of $\overline{N}$, $\overline{X}$ and $\overline{R}$, Case 4, for $\overline{p}^{(l)}(i)$ and $\overline{p}^{(l)}$, in Example 16.2.1

α	$\overline{N}_2[\alpha]$	$\overline{X}_2[\alpha]$	$\overline{R}_2[\alpha]$
1	3.6737	0.1996	18.4108
0	[2.8494, 3.9200]	[0.0951, 0.3000]	[9.4980, 41.2198]

Table 16.7: Alpha Cuts of $\overline{N}$, $\overline{X}$ and $\overline{R}$, Case 1, for $\overline{p}^{(l)}(i)$ and $\overline{p}^{(l)}$, in Example 16.2.1

Step 4

Given a value of μ_1 and μ_2, whose sum is one, we can now compute $\overline{Z} = \mu_1 \overline{R}^{(N)} + \mu_2 \overline{R}^{(L)}$. Then we may determine m_z (where the membership value of $\overline{Z}$ is one) and L_z and R_z. Recall that we are approximating these triangular shaped fuzzy numbers by triangular fuzzy numbers, since we have only the $\alpha = 0$ and $\alpha = 1$ cuts, so these areas are just areas of right triangles of height one. The results are in Tables 16.8 for selected values of the μ_i.

Step 5

Using the results in Table 16.8 we may now easily find the values of the objective function, given in equation (16.1), for certain values of the λ_i. We can see from Table 16.8 that we may choose K_1 in equation (16.1) to be eight and the factor $(K_1 - L_z)$ will always be positive. Table 16.9 shows the values of the objective function for three selections of the λ_i. The notation (a, b, c) at the top of this table means $\lambda_1 = a$, $\lambda_2 = b$ and $\lambda_3 = c$.

We see from Table 16.9 that Case 3 ($c = 2$, $M = 4$) is best for all choices of the μ_i and λ_i. Recall that for given values of the parameters μ_i and λ_i we are looking for the case that produces the smallest value in Table 16.7. The optimal case has very small system capacity ($M = 4$) and then we would also need to look at the expected number of lost customers per time period (next chapter).

16.2.1   Ranking the Fuzzy Numbers

Here we want to minimize $\overline{Z}$. So let $\overline{Z}^i = \mu_1 \overline{R}^{(N)} + \mu_2 \overline{R}^{(L)}$ for $i = 1, 2, 3, 4$ where the values of $\overline{R}^{(N)}$ and $\overline{R}^{(L)}$ are determined as in Case i in Table 16.3.

Case	(μ_1, μ_2)	L_z	m_z	R_z
1	$(0.7, 0.3)$	2.0585	12.4315	4.6049
	$(0.5, 0.5)$	2.7436	14.1399	6.5477
	$(0.3, 0.7)$	3.4287	15.8483	8.4904
2	$(0.7, 0.3)$	4.8015	31.9078	10.5937
	$(0.5, 0.5)$	6.3152	36.6002	14.8622
	$(0.3, 0.7)$	7.8290	41.2928	19.1308
3	$(0.7, 0.3)$	1.2103	5.7380	3.5911
	$(0.5, 0.5)$	1.6156	6.3766	5.3664
	$(0.3, 0.7)$	2.0209	7.0153	7.1417
4	$(0.7, 0.3)$	3.4470	14.5696	8.1571
	$(0.5, 0.5)$	4.8413	16.1241	12.1677
	$(0.3, 0.7)$	6.2356	17.6786	16.1783

Table 16.8: Central Value and Certain Areas Under the Graph of $\overline{Z}$, the Four Cases in Example 16.2.1

Case	(μ_1, μ_2)	$(1/3, 1/3, 1/3)$	$(0.4, 0.4, 0.2)$	$(0.3, 0.5, 0.2)$
1	$(0.7, 0.3)$	7.6593	8.2702	8.9192
	$(0.5, 0.5)$	8.6480	9.0680	9.9564
	$(0.3, 0.7)$	9.6366	9.8659	10.9936
2	$(0.7, 0.3)$	15.2333	16.1612	19.0322
	$(0.5, 0.5)$	17.7157	18.2864	21.7780
	$(0.3, 0.7)$	20.1982	20.4117	24.5238
3	$(0.7, 0.3)$	5.3729	5.7293	5.6241
	$(0.5, 0.5)$	6.0425	6.1777	6.1769
	$(0.3, 0.7)$	6.7120	6.6261	6.7297
4	$(0.7, 0.3)$	9.0933	9.2805	10.2822
	$(0.5, 0.5)$	10.4835	10.1467	11.4432
	$(0.3, 0.7)$	11.8737	11.0128	12.6042

Table 16.9: Values of the Objective Function in Example 16.2.1 for Certain Values of the Parameters μ_i and λ_i

$\overline{Z}$	Triangular Fuzzy Number
Case 1	(8.66/14.14/27.24)
Case 2	(23.97/36.60/66.32)
Case 3	(3.15/6.38/17.11)
Case 4	(6.44/16.12/40.46)

Table 16.10: The Triangular Fuzzy Numbers for the $\overline{Z}$ in Example 16.2.1.1

For example, if $i = 3$ we use the values for $\overline{R}^{(N)}$ and $\overline{R}^{(L)}$ computed in Case 3 in the above equation to obtain $\overline{Z}^3$. Given values for the μ_i we then rank the fuzzy numbers $\overline{Z}^i$ from smallest to largest and pick out those in the smallest set H_1 for the optimal solution.

Example 16.2.1.1

Here we will use only $\mu_1 = \mu_2 = 0.5$. From Table 16.3 we computed the triangular fuzzy number approximations to the $\overline{R}_i$, $1 \le i \le 8$ and then for $\overline{Z}^j$, for $j = 1, 2, 3, 4$. See previous examples on the ranking of fuzzy sets about approximating the triangular shaped fuzzy number $\overline{Z}^j$ by a triangular fuzzy number.

We first round off the $\alpha = 0$ cut and the $\alpha = 1$ cut of $\overline{R}_i$, $1 \le i \le 8$, to two decimal places for ease in producing a graph of these fuzzy numbers. Then we found the triangular fuzzy number approximations to $\overline{Z}^j$. These triangular fuzzy numbers for the $\overline{Z}^j$ are shown in Table 16.10.

Next we graph these triangular fuzzy number from Table 16.10 to see how they line up from smallest to largest. This is now shown in Figure 16.1. We want H_1 the set of smallest fuzzy numbers. From Figure 16.1 H_1 consists of $\overline{Z}^3$, $H_2 = \{\overline{Z}^1, \overline{Z}^4\}$ and $H_3 = \{\overline{Z}^2\}$. The optimal solution is Case 3, the same result as in Example 16.2.1. These results are easily seen by comparing the height of intersection with the horizontal line through 0.8 on the $y-axis$.

16.3 Fuzzy Arrival/Service Rates

We will have two values of $\overline{\lambda}$: (1) $\overline{\lambda}^{(n)}$ for the normal ; and (2) $\overline{\lambda}^{(l)}$ for the "long" customers. We will also have two values for $\overline{\mu}$: (1) $\overline{\mu}^{(n)}$ for our normal customers; and (2) $\overline{\mu}^{(l)}$ for those who take much longer in a server. We could use the "ideal point" method presented in the previous Chapter, but instead we will employ the optimization techniques from Chapters 13 and 14. We also only concentrate on minimizing $\overline{R}$.

For a value of c and M let $\overline{R} = \overline{R}^{(N)}$ when we use $\overline{\lambda}^{(n)}$ and $\overline{\mu}^{(n)}$, and for

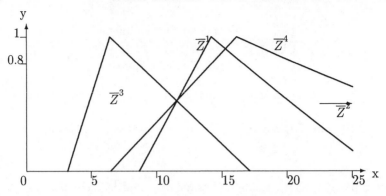

Figure 16.1: Ranking the Fuzzy Numbers $\overline{Z}^j$ in Example 16.2.1.1

the same values of c and M let $\overline{R} = \overline{R}^{(L)}$ when we have $\overline{\lambda}^{(l)}$ and $\overline{\mu}^{(l)}$. We form our compromise goal $\overline{Z} = \mu_1 \overline{R}^{(N)} + \mu_2 \overline{R}^{(L)}$. Our objective would be to $min\overline{Z}$ for $c = 1, 2, 3$ and $M = 4, 5, ..., 10$.

Example 16.3.1

We choose for the normal fuzzy arrival rate $\overline{\lambda}^{(n)} = (4/5/6)$ and the normal fuzzy service rate $\overline{\mu}^{(n)} = (5/6/7)$. These two rates were used together in Example 14.3.1 in Chapter 14. With respect to the long tailed distribution we picked for the fuzzy arrival rate $\overline{\lambda}^{(l)} = (3/4/5)$ and the fuzzy service rate will be $\overline{\mu}^{(l)} = (2/3/4)$. Notice that $\overline{\lambda}^{(l)} \approx 0.8\overline{\lambda}^{(n)}$ and $\overline{\mu}^{(l)} \approx 0.5\overline{\mu}^{(n)}$. All these fuzzy numbers would be obtained from statistical data as explained in Chapter 3.

We will consider four cases shown in Table 16.11 These are like the cases used in Example 16.2.1. Notice that in each case we : (1) first use the given value of c and M, $\overline{\mu}^{(n)} = (5/6/7)$ and $\overline{\lambda}^{(n)} = (4/5/6)$ to compute α-cuts of $\overline{R}^{(N)}$; and (2) next use the same values of c, M and $\overline{\mu}^{(l)} = (2/3/4)$ and $\overline{\lambda}^{(l)} = (3/4/5)$ to obtain α-cuts of $\overline{R}^{(L)}$.

We will number the fuzzy numbers $\overline{U}_i$, $\overline{N}_i$, $\overline{X}_i$ and $\overline{R}_i$ as follows: (1) $i = 1, 3, 5, 7$ for Cases 1,2,3,4 when we use $\overline{\lambda}^{(n)}$ and $\overline{\mu}^{(n)}$; and (2) $i = 2, 4, 6, 8$ for Cases 1,2,3,4 when we have $\overline{\lambda}^{(l)}$ and $\overline{\mu}^{(l)}$. Now we go through four steps: (1) find the needed fuzzy steady state probabilities in Step 1; (2) Step 2 determines the $\overline{U}_i, ..., \overline{R}_i$, $1 \le i \le 8$; (3) Step 3 computes $\overline{Z} = \mu_1 \overline{R}^{(N)} + \mu_2 \overline{R}^{(L)}$, for selected values of the μ_i and then finds the central value of $\overline{Z}$, called m_z, and computes L_z (R_z) the area under the graph of the membership function to the left (right) of m_z; and (4) step 4 solves the minimization problem

$$min\{\lambda_1(K_1 - L_z) + \lambda_2 m_z + \lambda_3 R_z\}, \tag{16.2}$$

Case	Number of Servers	System Capacity	$\overline{\lambda}$	$\overline{\mu}$	$\overline{R}$
1	$c = 1$	$M = 4$	$\overline{\lambda}^{(n)}$	$\overline{\mu}^{(n)}$	$\overline{R}^{(N)}$
			$\overline{\lambda}^{(l)}$	$\overline{\mu}^{(l)}$	$\overline{R}^{(L)}$
2	$c = 1$	$M = 10$	$\overline{\lambda}^{(n)}$	$\overline{\mu}^{(n)}$	$\overline{R}^{(N)}$
			$\overline{\lambda}^{(l)}$	$\overline{\mu}^{(l)}$	$\overline{R}^{(L)}$
3	$c = 2$	$M = 4$	$\overline{\lambda}^{(n)}$	$\overline{\mu}^{(n)}$	$\overline{R}^{(N)}$
			$\overline{\lambda}^{(l)}$	$\overline{\mu}^{(l)}$	$\overline{R}^{(L)}$
4	$c = 2$	$M = 10$	$\overline{\lambda}^{(n)}$	$\overline{\mu}^{(n)}$	$\overline{R}^{(N)}$
			$\overline{\lambda}^{(l)}$	$\overline{\mu}^{(l)}$	$\overline{R}^{(L)}$

Table 16.11: The Four Cases in Example 16.3.1

the same as equation (13.19), for selected values of the λ_i.

Step 1

We calculated the alpha-cuts $\overline{w}_i[\alpha]$ of the fuzzy steady state probabilities, for $\alpha = 0, 1/3, 2/3, 1$, for Cases 1,2,3,4 using $\overline{\lambda}^{(n)}$ and $\overline{\mu}^{(n)}$, in Example 14.3.1 in Chapter 14. Therefore, we only need to do this using $\overline{\lambda}^{(l)}$ and $\overline{\mu}^{(l)}$. We did this using the Premium Solver Platform V5.0 from Frontline Systems [1] as explained after Example 12.2.1 in Chapter 12. Table 16.12 presents the $\alpha = 0$ cut for Case 4 using $\overline{\lambda}^{(l)}$ and $\overline{\mu}^{(l)}$ while Table 16.13 is for the alpha zero cut in Case 1 when we employ $\overline{\lambda}^{(l)}$ and $\overline{\mu}^{(l)}$. The 0.0000* in Table 16.12 means it is 0.0000 rounded off to four decimal places.

Step 2

Here we are to get the alpha-cuts, $\alpha = 0, 1/3, 2/3$, of the $\overline{U}_i$, $\overline{N}_i$, $\overline{X}_i$ and $\overline{R}_i$, $1 \leq i \leq 8$. We will not need the $\overline{U}_i$ in this example but it can be used to obtain $\overline{X}_i$. We obtained these alpha-cuts using $\overline{\lambda}^{(n)}$ and $\overline{\mu}^{(n)}$ in Example 14.3.1, although we only used those for $\overline{N}_i$ in that example, for $i = 1, 3, 5, 7$. So we need to do this for $\overline{\lambda}^{(l)}$ and $\overline{\mu}^{(l)}$ for $i = 2, 4, 6, 8$. In all cases we find the alpha-cuts of $\overline{U}_i$, $\overline{N}_i$ and $\overline{X}_i$ by solving a linear programming problem as discussed in previous chapters, so we do not present these details again. $\overline{R}$ is just $\overline{N}/\overline{X}$ evaluated using interval arithmetic (Chapter 2).

Table 16.14 shows the results for $\overline{N}$, $\overline{X}$ and $\overline{R}$ for Case 4 for $\overline{\lambda}^{(l)}$ and $\overline{\mu}^{(l)}$ and Table 13.15 gives the alpha-cuts of these fuzzy numbers in Case 1 when we are using $\overline{\lambda}^{(l)}$ and $\overline{\mu}^{(l)}$.

$\overline{w}$	Alpha Zero Cut
$\overline{w}_0[0]$	$[0.0119, 0.4546]$
$\overline{w}_1[0]$	$[0.0297, 0.3432]$
$\overline{w}_2[0]$	$[0.0371, 0.1816]$
$\overline{w}_3[0]$	$[0.0464, 0.1268]$
$\overline{w}_4[0]$	$[0.0180, 0.1036]$
$\overline{w}_5[0]$	$[0.0067, 0.0955]$
$\overline{w}_6[0]$	$[0.0025, 0.0983]$
$\overline{w}_7[0]$	$[0.0009, 0.1135]$
$\overline{w}_8[0]$	$[0.0004, 0.1417]$
$\overline{w}_9[0]$	$[0.0001, 0.1771]$
$\overline{w}_{10}[0]$	$[0.0000^*, 0.2214]$

Table 16.12: Alpha Zero Cuts of the Fuzzy Steady State Probabilities, Case 4, $\overline{\lambda}^{(l)}$, $\overline{\mu}^{(l)}$, in Example 16.3.1

$\overline{w}$	Alpha Zero Cut
$\overline{w}_0[0]$	$[0.0155, 0.3278]$
$\overline{w}_1[0]$	$[0.0388, 0.2458]$
$\overline{w}_2[0]$	$[0.0970, 0.2000]$
$\overline{w}_3[0]$	$[0.1383, 0.2608]$
$\overline{w}_4[0]$	$[0.1037, 0.6062]$

Table 16.13: Alpha Zero Cuts of the Fuzzy Steady State Probabilities, Case 1, $\overline{\lambda}^{(l)}$, $\overline{\mu}^{(l)}$, in Example 16.3.1

α	$\overline{N}_8[\alpha]$	$\overline{X}_8[\alpha]$	$\overline{R}_8[\alpha]$
1	2.2503	3.9720	0.5665
2/3	$[1.5583, 3.3316]$	$[2.9294, 5.1326]$	$[0.3036, 1.1373]$
1/3	$[1.1434, 5.1429]$	$[2.1205, 6.6353]$	$[0.1723, 2.4253]$
0	$[0.8678, 7.1466]$	$[1.4952, 7.7860]$	$[0.1114, 4.7797]$

Table 16.14: Alpha Cuts of $\overline{N}$, $\overline{X}$ and $\overline{R}$, Case 4, for $\overline{\lambda}^{(l)}$ and $\overline{\mu}^{(l)}$, in Example 16.3.1

α	$\overline{N}_2[\alpha]$	$\overline{X}_2[\alpha]$	$\overline{R}_2[\alpha]$
1	2.5557	2.6889	0.9505
2/3	[2.1898, 2.8839]	[2.2299, 3.1316]	[0.6993, 1.2933]
1/3	[1.8102, 3.1615]	[1.7738, 3.5486]	[0.5101, 1.7823]
0	[1.4443, 3.3851]	[1.3444, 3.9380]	[0.3668, 2.5179]

Table 16.15: Alpha Cuts of $\overline{N}$, $\overline{X}$ and $\overline{R}$, Case 1, for $\overline{\lambda}^{(l)}$ and $\overline{\mu}^{(l)}$, in Example 16.3.1

Case	(μ_1, μ_2)	L_z	m_z	R_z
1	(0.7, 0.3)	0.1818	0.5504	0.3462
	(0.5, 0.5)	0.2235	0.6648	0.4338
	(0.3, 0.7)	0.2652	0.7790	0.5214
2	(0.7, 0.3)	0.5200	1.2348	0.9600
	(0.5, 0.5)	0.6695	1.6054	1.1302
	(0.3, 0.7)	0.8190	1.9761	1.1302
3	(0.7, 0.3)	0.0981	0.2598	0.2461
	(0.5, 0.5)	0.1212	0.3067	0.3261
	(0.3, 0.7)	0.1444	0.3536	0.4060
4	(0.7, 0.3)	0.1368	0.3109	0.5770
	(0.5, 0.5)	0.1820	0.3839	0.8442
	(0.3, 0.7)	0.2272	0.4569	1.1113

Table 16.16: Central Value and Certain Areas Under the Graph of $\overline{Z}$, the Four Cases in Example 16.3.1

Step 3

Given a value of μ_1 and μ_2, whose sum is one, we can now compute $\overline{Z} = \mu_1 \overline{R}^{(N)} + \mu_2 \overline{R}^{(L)}$. Then we may determine m_z (where the membership value of $\overline{Z}$ is one) and L_z and R_z using the Trapezoidal Rule discussed in Example 13.3.1.1 in Chapter 13. Since we have only an approximation to the membership functions for $\overline{R}^{(N)}$ and $\overline{R}^{(L)}$ at $\alpha = 0, 1/3, 2/3, 1$, L_z and R_z will be only approximations to the true areas. The results are in Table 16.16 for selected values of the μ_i.

Step 4

Using the results in Table 16.16 we may now easily find the values of the objective function, given in equation (16.2), for certain values of the λ_i. We can see from Table 16.16 that we may choose K_1 in equation (16.2) to be

Case	(μ_1, μ_2)	$(1/3.1/3.1/3)$	$(0.4, 0.4, 0.2)$	$(0.3, 0.5, 0.2)$
1	$(0.7, 0.3)$	0.5716	0.6167	0.5899
	$(0.5, 0.5)$	0.6250	0.6632	0.6521
	$(0.3, 0.7)$	0.6784	0.7098	0.7142
2	$(0.7, 0.3)$	0.8916	0.8779	0.9534
	$(0.5, 0.5)$	1.0220	1.0004	1.1279
	$(0.3, 0.7)$	1.1524	1.1229	1.3024
3	$(0.7, 0.3)$	0.4693	0.5139	0.4497
	$(0.5, 0.5)$	0.5038	0.5394	0.4822
	$(0.3, 0.7)$	0.5384	0.5649	0.5147
4	$(0.7, 0.3)$	0.5837	0.5850	0.5298
	$(0.5, 0.5)$	0.6820	0.6496	0.6062
	$(0.3, 0.7)$	0.7804	0.7142	0.6826

Table 16.17: Values of the Objective Function in Example 16.3.1 for Certain Values of the Parameters μ_i and λ_i

one and the factor $(K_1 - L_z)$ will always be positive. Table 16.17 shows the values of the objective function for three selections of the λ_i. The notation (a, b, c) at the top of this tables means $\lambda_1 = a$, $\lambda_2 = b$ and $\lambda_3 = c$.

We see from Table 16.17 that Case 3 ($c = 2$, $M = 4$) is best for all choices of the μ_i and λ_i. Recall that for given values of the parameters μ_i and λ_i we are looking for the case that produces the smallest value in Table 16.17. The optimal case has very small system capacity ($M = 4$) we would also need to look at the expected number of lost customers per time period.

16.3.1   Ranking the Fuzzy Numbers

Here we want to minimize $\overline{Z}$. So let $\overline{Z}^i = \mu_1 \overline{R}^{(N)} + \mu_2 \overline{R}^{(L)}$ for $i = 1, 2, 3, 4$ where the values of $\overline{R}^{(N)}$ and $\overline{R}^{(L)}$ are determined as in Case i in Table 16.11. For example, if $i = 3$ we use the values for $\overline{R}^{(N)}$ and $\overline{R}^{(L)}$ computed in Case 3 in the above equation to obtain $\overline{Z}^3$. Given values for the μ_i we then rank the fuzzy numbers $\overline{Z}^i$ from smallest to largest and pick out those in the smallest set H_1 for the optimal solution.

Example 16.3.1.1

Here we will use only $\mu_1 = \mu_2 = 0.5$. From Table 16.11 we computed the triangular fuzzy number approximations to the $\overline{R}_i$, $1 \le i \le 8$ and then for $\overline{Z}^j$, for $j = 1, 2, 3, 4$. See previous examples on the ranking of fuzzy sets about approximating the triangular shaped fuzzy number $\overline{Z}^i$ by a triangular fuzzy number.

$\overline{Z}$	Triangular Fuzzy Number
Case 1	(0.27/0.66/1.70)
Case 2	(0.40/1.60/4.32)
Case 3	(0.10/0.30/1.12)
Case 4	(0.10/0.38/2.71)

Table 16.18: The Triangular Fuzzy Numbers for the $\overline{Z}$ in Example 16.3.1.1

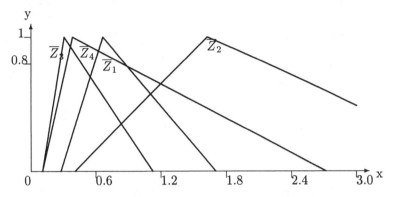

Figure 16.2: Ranking the Fuzzy Numbers $\overline{Z}^j$ in Example 16.3.1.1

We first round off the $\alpha = 0$ cut and the $\alpha = 1$ cut of $\overline{R}$, $1 \le i \le 8$, to two decimal places for ease in producing a graph of these fuzzy numbers. Then we found the triangular fuzzy number approximations to $\overline{Z}^j$. These triangular fuzzy numbers for the $\overline{Z}^j$ are shown in Table 16.18.

Next we graph these triangular fuzzy number from Table 16.18 to see how they line up from smallest to largest. This is now shown in Figure 16.2. We want H_1 the set of smallest fuzzy numbers. From Figure 16.2 H_1 consists of $\overline{Z}^3$ and $\overline{Z}^4$, $H_2 = \{\overline{Z}^1\}$ and $H_3 = \{\overline{Z}^2\}$. The optimal solution is Cases 3 and 4. We note, using the height of intersection rule, that $\overline{Z}^4 \approx \overline{Z}^3$, $\overline{Z}^4 \approx \overline{Z}^1$ but $\overline{Z}^3 < \overline{Z}^1$ so H_1 consists of both $\overline{Z}^3$ and $\overline{Z}^4$. The optimal solution was Case 3 in Example 16.3.1.

16.4   References

1. Frontline Systems (www.frontsys.com).

2. D.A.Menasce and V.A.F.Almeida: Capacity Planning for Web Performance, Prentice Hall, Upper Saddle River, N.J., 1998.

Chapter 17

Putting It All Together

17.1 Introduction

In the previous chapters, Chapters 13-16, we looked at various optimization strategies. We also considered two queuing models: (1) the first one based on fuzzy probabilities; (2) and the second one used fuzzy arrival and fuzzy service rates.

In Chapter 13 we studied the problem of minimizing $\overline{R}$, a fuzzy number representing average response time for the system, and maximizing $\overline{U}$, a fuzzy number for server utilization. In this chapter two of our fuzzy goals can be to $min\overline{R}$ and $max\overline{U}$. Another fuzzy goal we might consider is to minimize the number of lost customers per unit time $\overline{LC}$ in equation (13.16) in Chapter 13, due to having finite system capacity.

In Chapter 14 we had the problem of maximizing fuzzy profit subject to various constraints like number of servers, system capacity, different advertising rates, etc. The fuzzy number $\overline{N}$, representing the expected number of customers in the system, was needed in this model. In this chapter we will add the fuzzy goal of $max\overline{Profit}$ to the previous fuzzy goals, but the constraints will be slightly different. In this chapter we will use the notation $\overline{\Pi}$ for fuzzy profit.

In Chapter 15 we discussed a method of handling the problem of "burstiness". In this chapter we will continue to incorporate "burstiness" but we will not use "ideal" points as was done in Chapter 15.

In Chapter 16 we discussed the problem of "long tailed distributions" in queuing systems used in web systems. In this chapter we also involve the phenomena of "long tailed distributions" in our optimization models.

We first consider the queuing model based on fuzzy probabilities and then the model derived from fuzzy arrival (service) rates. In each case we will only develop one example having multiple fuzzy goals incorporating both "burstiness" and "long tailed distributions". After studying Chapters 13-16 the reader may easily develop other optimization problems involving multiple fuzzy goals. We will not use the method of "ranking the fuzzy sets" in this chapter because we have conflicting goals and we will end up, as in Examples 13.2.3.2 and 13.3.3.1, with conflicting optimal solutions.

17.2   Fuzzy Probabilities

We first specify "normal" operating conditions. Under normal operating conditions the fuzzy arrival probabilities are $\overline{p}^{(n)}(i)$ and the normal fuzzy service probability is $\overline{p}^{(n)}$. For a given number of servers (c) and system capacity (M) and using $\overline{p}^{(n)}(i)$ and $\overline{p}^{(n)}$ we compute fuzzy numbers for $\overline{U}$, $\overline{N}$, $\overline{R}$, $\overline{LC}$ and $\overline{\Pi}$. For notational convience we will rewrite these fuzzy numbers as $\overline{U}^{(N)}$, $\overline{N}^{(N)}$, $\overline{R}^{(N)}$, $\overline{LC}^{(N)}$ and $\overline{\Pi}^{(N)}$.

The expression for fuzzy profit, equation (14.1) of Chapter 14, is changed to

$$\overline{\Pi} = \overline{T}\ \overline{N} - [\overline{Q}(M - c) + \overline{C}c], \qquad (17.1)$$

because now we are not allowing for advertising rates to effect arrival rates and we will have only one type of server at cost $\overline{C}$ dollars per unit time.

Burstiness effects the arrival probabilities and let the burstiness fuzzy arrival probability be $\overline{p}^{(b)}(i)$. For a given value of c, M and employing $\overline{p}^{(b)}(i)$ and $\overline{p}^{(n)}$ we determine the fuzzy numbers $\overline{U}^{(B)}$, $\overline{N}^{(B)}$, $\overline{R}^{(B)}$, $\overline{LC}^{(B)}$ and $\overline{\Pi}^{(B)}$. The superscript relates to "burstiness".

Long tailed distributions means there are different customers with their fuzzy arrival probabilities $\overline{p}^{(l)}(i)$ and their fuzzy service probability $\overline{p}^{(l)}$. Using c, M and $\overline{p}^{(l)}(i)$ and $\overline{p}^{(l)}$ we also compute the fuzzy numbers $\overline{U}^{(L)}$, $\overline{N}^{(L)}$, $\overline{R}^{(L)}$, $\overline{LC}^{(L)}$ and $\overline{\Pi}^{(L)}$ with the superscript L for "long tailed distributions".

We have the three fuzzy goals :(1) $min\overline{R}$; (2) $min\overline{LC}$; and (3) $max\overline{\Pi}$. So we now need to get $\overline{R}$, $\overline{LC}$ and $\overline{\Pi}$ from those computed for "normal", "burstiness" and "long tailed distributions". From previous data we estimate that: (1) normal operating conditions occur approximately 84% of the time; (2) burstiness is about 10% of the time; and (3) long tailed distributions happens around 6% of operating time. Now these percentages have all been estimated from data and so they all become fuzzy numbers (see Chapter 3). Let $\overline{\tau}_1 = (0.81/0.84/0.87)$, $\overline{\tau}_2 = (0.07/0.10/0.13)$ and $\overline{\tau}_3 = (0.03/0.06/0.09)$. Then $\overline{\tau}_1 \approx 84\%$ is for normal operations, $\overline{\tau}_2 \approx 10\%$ goes with burstiness and $\overline{\tau}_3 \approx 6\%$ we associate with long tailed distribution conditions.

Then we calculate the fuzzy number for $\overline{R}$ as

$$\overline{R} = \overline{\tau}_1\overline{R}^{(N)} + \overline{\tau}_2\overline{R}^{(B)} + \overline{\tau}_3\overline{R}^{(L)}, \qquad (17.2)$$

and for $\overline{LC}$

$$\overline{LC} = \overline{\tau}_1\overline{LC}^{(N)} + \overline{\tau}_2\overline{LC}^{(B)} + \overline{\tau}_3\overline{LC}^{(L)}, \qquad (17.3)$$

and $\overline{\Pi}$

$$\overline{\Pi} = \overline{\tau}_1 \overline{\Pi}^{(N)} + \overline{\tau}_2 \overline{\Pi}^{(B)} + \overline{\tau}_3 \overline{\Pi}^{(L)}. \qquad (17.4)$$

These are all evaluated using interval arithmetic (Chapter 2). This is easily done when all the fuzzy numbers are positive. However, fuzzy profit need not be strictly positive ($\overline{\Pi} > 0$) since the left end point of the $\alpha = 0$ cut can be negative. Then we need to use equation (2.12) in Chapter 2 to find the product $\overline{\tau}_3[\alpha]\overline{\Pi}^{(W)}[\alpha]$ for $W = N$, or $W = B$ or $W = L$.

We now combine our three fuzzy goals in to a single fuzzy goal of $max\overline{Z}$ where

$$\overline{Z} = \mu_1 \overline{\Pi} + \mu_2[K_1 - \overline{LC}] + \mu_3[K_2 - \overline{R}], \qquad (17.5)$$

where $\mu_i > 0$ all i, $\mu_1 + \mu_2 + \mu_3 = 1$ and the K_i are sufficiently large positive constants so that $min\overline{LC}$ is equivalent to $max[K_1 - \overline{LC}]$ and $min\overline{R}$ is the same as $max[K_2 - \overline{R}]$. Of course, we would choose different values for the μ_i and present the results to management. For example, if we want to emphasize profit first, and then lost customers, we might choose $\mu_1 = 0.5$, $\mu_2 = 0.3$ and $\mu_3 = 0.2$.

Now we employ the techniques discussed in Section 2.5 of Chapter 2, also discussed in Chapter 14, to $max\overline{Z}$. Let m_z be the central value (where the membership function value is one)of $\overline{Z}$, and L_z (R_z) the area under the graph of the membership function for $\overline{Z}$ to the left (right) of m_z. Finally, we solve

$$max(\lambda_1[K_3 - L_z] + \lambda_2 m_z + \lambda_3 R_z), \qquad (17.6)$$

for $\lambda_i > 0$, all i, $\lambda_1 + \lambda_2 + \lambda_3 = 1$, where K_3 is a sufficiently large positive constant so that $minL_z$ is the same as $max[K_3 - L_z]$, usually we can choose $K_3 = 1$. Again we choose various values for the λ_i to show our different emphasis on the three goals of $minL_z$, $maxm_z$ and $maxR_z$.

Example 17.2.1

There will be only four cases to consider for an optimal design: (1) Case 1 is $c = 1$ and $M = 4$; (2) Case 2 has $c = 1$, $M = 10$; (3) Case 3 will be $c = 2$ with $M = 4$; and (4) $c = 2$, $M = 10$ comprises Case 4. These are the values of c and M that we used in Chapter 13-16.

Here we use values for c, M and the fuzzy arrival (service) probabilities, to compute the $\overline{U}^{(N)}, ..., \overline{\Pi}^{(L)}$ and then find the values for $\overline{R}, \overline{LC}$ and $\overline{\Pi}$.

Now we pick values for the fuzzy arrival (service) probabilities. First we set $\overline{p}^{(n)}(i)$ to be given in Table 6.1 and $\overline{p}^{(n)} = (0.3/0.4/0.5)$. These two values were used together in Chapter 13. Next $\overline{p}^{(b)}(i)$ is in Table 15.1 and lastly $\overline{p}^{(l)}(i)$ is defined in Table 16.1, $\overline{p}^{(l)} = (0.1/0.2/0.3)$ both from Chapter 16.

All the needed fuzzy steady state probabilities have been determined in previous chapters. In fact, all the needed fuzzy numbers $\overline{U}^{(N)}, ..., \overline{R}^{(L)}$, without the $\overline{\Pi}$ and $\overline{LC}$ values, were found in previous chapters. So now we first

Case	Alpha Equal Zero	Alpha Equal One
$\overline{p}^{(n)}(i)$	$[1.83, 3.27]$	2.55
$\overline{p}^{(b)}(i)$	$[1.43, 2.37]$	1.90
$\overline{p}^{(l)}(i)$	$[0.4, 0.6]$	0.5

Table 17.1: $\alpha = 0, 1$ cuts of $\overline{\Lambda}$ in Equation (17.7) for the Three Cases of $\overline{p}(i)$ in Example 17.2.1

need to find $\overline{\Pi}^{(W)}$, $W = N, B, L$ from equation (17.1). In equation (17.1) we use $\overline{Q} = (0.04/0.07/0.10)$, $\overline{C} = (0.27/0.30/0.33)$ and $\overline{T} = (0.83/0.95/1.07)$ as were employed in Chapter 14. To obtain $\overline{LC}$ in equation (13.16) we need $\overline{w}_M$, which was found from the fuzzy steady state probabilities, and

$$\overline{\Lambda} = \sum_{i=0}^{L} i\overline{p}(i), \tag{17.7}$$

using $\overline{p}^{(n)}(i)$, $\overline{p}^{(b)}(i)$ and $\overline{p}^{(l)}(i)$ for $\overline{p}(i)$. We did this and the $\alpha = 0, 1$ cuts of the results are in Table 17.1. Then we just multiply $\overline{w}_M[\alpha]$ and the required result from Table 17.1 to get the $\alpha = 0, 1$ cut of $\overline{LC}$.

Having determined the fuzzy numbers for fuzzy profit and the expected number of lost customers per unit time, we may next get the fuzzy numbers for $\overline{R}$, $\overline{LC}$ and $\overline{\Pi}$ from equations (17.2) to (17.4). All fuzzy numbers are approximated through only two alpha-cuts $\alpha = 0, 1$, so equations (17.2) to (17.4) are evaluated using interval arithmetic for $\alpha = 0$.

The next step is to get $\overline{Z}$ from equation (17.5). In this calculation we used $K_1 = 4$ and $K_2 = 20$. It turns out that in a couple of cases $K_2 - \overline{R}$ is negative. This in turn can make m_z in Table 17.2 negative (Case 2, $(0.2, 0.3, 0.5)$). We usually pick the K-value so that the term $[K - (*)]$ is positive, but it does not matter because it only effects the value of the objective function but not the optimal solutions.

From $\overline{Z}$ we obtain L_z, m_z and R_z and these numbers are presented in Table 17.2 for various choices for the μ_i.

Using the results in Table 17.2 we may now easily find the values of the objective function, given in equation (17.6), for certain values of the λ_i. We used $K_3 = 10$ in these calculations. Table 17.3 shows the values of the objective function for three selections of the λ_i. The notation (a, b, c) at the top of this tables means $\lambda_1 = a$, $\lambda_2 = b$ and $\lambda_3 = c$. What is the optimal solution? Recall that this is a maximization problem.

We now obtain some interesting mixed results. First consider the case where we weight the three goals equally ($\lambda_i = 1/3$ all i). From Table 17.3 we find that Case 3 ($c = 2, M = 4$) is the best for all choices of the μ_i (equation (17.5)). When $(\mu_1, \mu_2, \mu_3) = (0.3, 0.5, 0.2)$ Case 4 gives a close alternate solution.

Case	(μ_1, μ_2, μ_3)	L_z	m_z	R_z
1	$(0.5, 0.2, 0.3)$	1.2949	4.8505	0.8086
	$(0.3, 0.5, 0.2)$	1.0115	3.7723	0.6678
	$(0.2, 0.3, 0.5)$	1.8658	5.9835	1.0237
2	$(0.5, 0.2, 0.3)$	3.0220	2.7217	1.8892
	$(0.3, 0.5, 0.2)$	2.1412	2.1772	1.3638
	$(0.2, 0.3, 0.5)$	4.3314	-0.9082	2.3633
3	$(0.5, 0.2, 0.3)$	0.9645	6.3265	0.6058
	$(0.3, 0.5, 0.2)$	0.8063	4.8681	0.5422
	$(0.2, 0.3, 0.5)$	1.2858	8.6930	0.6681
4	$(0.5, 0.2, 0.3)$	2.1322	6.6123	1.3599
	$(0.3, 0.5, 0.2)$	1.5620	4.8874	1.0195
	$(0.2, 0.3, 0.5)$	2.7884	5.8712	1.4414

Table 17.2: Central Value and Certain Areas Under the Graph of $\overline{Z}$, the Four Cases in Example 17.2.1

Next look at $(\lambda_1, \lambda_2, \lambda_3) = (0.2, 0.4, 0.4)$ and we see that: (1) if $(\mu_1, \mu_2, \mu_3) = (0.5, 0.2, 0.3)$, then Case 4 is best; (2) if $(\mu_1, \mu_2, \mu_3) = (0.3, 0.5, 0.2)$, then Cases 3 and 4 are approximately equal; and (3) if $(\mu_1, \mu_2, \mu_3) = (0.2, 0.3, 0.5)$, then Case 3 produces the maximum value.

Finally we have $(\lambda_1, \lambda_2, \lambda_3) = (0.2, 0.5, 0.3)$. The results are identical to those in the previous case (paragraph).

Obviously, the the final results depend heavily on the values of the weights (μ_i, λ_i).

17.3 Fuzzy Arrival/Service Rates

We first specify "normal" operating conditions. Under normal operating conditions the fuzzy arrival rate is $\overline{\lambda}^{(n)}$ and the normal fuzzy service rate is $\overline{\mu}^{(n)}$. For a given number of servers (c) and system capacity (M) and using $\overline{\lambda}^{(n)}$ and $\overline{\mu}^{(n)}$ we compute fuzzy numbers for $\overline{U}, \overline{N}, \overline{R}$ and $\overline{\Pi}$. For notational convience we will rewrite these fuzzy numbers as $\overline{U}^{(N)}, \overline{N}^{(N)}, \overline{R}^{(N)}$ and $\overline{\Pi}^{(N)}$.

The expression for fuzzy profit, equation (14.1) of Chapter 14, is now changed to

$$\overline{\Pi} = \overline{T}\,\overline{N} - [\overline{Q}(M - c) + \overline{C}c], \qquad (17.8)$$

because now we are not allowing for advertising rates to effect arrival rates and we will have only one type of server at cost $\overline{C}$ dollars per unit time.

Burstiness effects the arrival rate and let the burstiness fuzzy arrival rate be $\overline{\lambda}^{(b)}$. The fuzzy arrival rate for burstiness will be $\overline{\lambda}^{(n)}$ shifted to the right.

Case	(μ_1, μ_2, μ_3)	$(1/3, 1/3, 1/3)$	$(0.2, 0.4, 0.4)$	$(0.2, 0.5, 0.3)$
1	$(0.5, 0.2, 0.3)$	4.7881	4.0047	4.4088
	$(0.3, 0.5, 0.2)$	4.4762	3.5738	3.8842
	$(0.2, 0.3, 0.5)$	5.0471	4.4297	4.9257
2	$(0.5, 0.2, 0.3)$	3.8630	3.2400	3.3232
	$(0.3, 0.5, 0.2)$	3.7999	2.9882	3.0695
	$(0.2, 0.3, 0.5)$	2.3746	1.7158	1.3886
3	$(0.5, 0.2, 0.3)$	5.3226	4.5800	5.1521
	$(0.3, 0.5, 0.2)$	4.8680	4.0029	4.4354
	$(0.2, 0.3, 0.5)$	6.0251	5.4873	6.2898
4	$(0.5, 0.2, 0.3)$	5.2800	4.7625	5.2877
	$(0.3, 0.5, 0.2)$	4.7816	4.0504	4.4371
	$(0.2, 0.3, 0.5)$	4.8413	4.3672	4.8102

Table 17.3: Values of the Objective Function in Example 17.2.1 for Certain Values of the Parameters μ_i and λ_i

For a given value of c, M and employing $\overline{\lambda}^{(b)}$ and $\overline{\mu}^{(n)}$ we determine the fuzzy numbers $\overline{U}^{(B)}$, $\overline{N}^{(B)}$, $\overline{R}^{(B)}$ and $\overline{\Pi}^{(B)}$. The superscript relates to "burstiness".

Long tailed distributions means there are different customers with their fuzzy arrival rate $\overline{\lambda}^{(l)}$ and their fuzzy service rate $\overline{\mu}^{(l)}$. Now $\overline{\lambda}^{(l)}$ will be $\overline{\lambda}^{(n)}$ shifted to the left and $\overline{\mu}^{(l)}$ is also $\overline{\mu}^{(n)}$ shifted to the left. Using c, M and $\overline{\lambda}^{(l)}$ and $\overline{\mu}^{(l)}$ we also compute the fuzzy numbers $\overline{U}^{(L)}$, $\overline{N}^{(L)}$, $\overline{R}^{(L)}$ and $\overline{\Pi}^{(L)}$ with the superscript L for "long tailed distributions".

We have the three fuzzy goals :(1) $min\overline{R}$; (2) $max\overline{U}$; and (3) $max\overline{\Pi}$. So we now need to get $\overline{R}$, $\overline{U}$ and $\overline{\Pi}$ from those computed for "normal", "burstiness" and "long tailed distributions". From previous data we estimate that: (1) normal operating conditions occur approximately 84% of the time; (2) burstiness is about 10% of the time; and (3) long tailed distributions happens around 6% of operating time. Now these percentages have all been estimated from data and so they all become fuzzy numbers (see Chapter 3). Let $\overline{\tau}_1 = (0.81/0.84/0.87)$, $\overline{\tau}_2 = (0.07/0.10/0.13)$ and $\overline{\tau}_3 = (0.03/0.06/0.09)$. Then $\overline{\tau}_1 \approx 84\%$ is for normal operations, $\overline{\tau}_2 \approx 10\%$ goes with burstiness and $\overline{\tau}_3 \approx 6\%$ we associate with long tailed distribution conditions.

Then we calculate the fuzzy number for $\overline{R}$ as

$$\overline{R} = \overline{\tau}_1 \overline{R}^{(N)} + \overline{\tau}_2 \overline{R}^{(B)} + \overline{\tau}_3 \overline{R}^{(L)}, \qquad (17.9)$$

and for $\overline{U}$

$$\overline{U} = \overline{\tau}_1 \overline{U}^{(N)} + \overline{\tau}_2 \overline{U}^{(B)} + \overline{\tau}_3 \overline{U}^{(L)}, \qquad (17.10)$$

and $\overline{\Pi}$

$$\overline{\Pi} = \overline{\tau}_1 \overline{\Pi}^{(N)} + \overline{\tau}_2 \overline{\Pi}^{(B)} + \overline{\tau}_3 \overline{\Pi}^{(L)}. \qquad (17.11)$$

These are all evaluated using interval arithmetic (Chapter 2). This is easily done when all the fuzzy numbers are positive. However, fuzzy profit need not be strictly positive ($\overline{\Pi} > 0$) since the left end point of the $\alpha = 0$, and maybe $\alpha = 1/3$, cut can be negative. Then we need to use equation (2.12) in Chapter 2 to find the product $\overline{\tau}_3[\alpha]\overline{\Pi}^{(W)}[\alpha]$ for $W = N$, or $W + B$ or $W = L$.

We now combine our three fuzzy goals in to a single fuzzy goal of $max\overline{Z}$ where

$$\overline{Z} = \mu_1\overline{\Pi} + \mu_2\overline{U} + \mu_3[K - \overline{R}], \qquad (17.12)$$

where $\mu_i > 0$ all i, $\mu_1 + \mu_2 + \mu_3 = 1$ and K is a sufficiently large positive constant so that $min\overline{R}$ is equivalent to $max[K - \overline{R}]$. Of course, we would choose different values for the μ_i and present the results to management. For example, if we want to emphasize profit first, and then server utilization, we might choose $\mu_1 = 0.5$, $\mu_2 = 0.3$ and $\mu_3 = 0.2$.

Now we employ the techniques discussed in Section 2.5 of Chapter 2, also discussed in Chapter 14, to $max\overline{Z}$. Let m_z be the central value (where the membership function value is one)of $\overline{Z}$, and L_z (R_z) the area under the graph of the membership function for $\overline{Z}$ to the left (right) of m_z. Finally, we solve

$$max(\lambda_1[K_1 - L_z] + \lambda_2 m_z + \lambda_3 R_z), \qquad (17.13)$$

for $\lambda_i > 0$, all i, $\lambda_1 + \lambda_2 + \lambda_3 = 1$, where K_1 is a sufficiently large positive constant so that $min L_z$ is the same as $max[K_1 - L_z]$, usually we can choose $K_1 = 1$. Again we choose various values for the λ_i to show our different emphasis on the three goals of $min L_z$, $max m_z$ and $max R_z$.

Example 17.3.1

There will be only four cases to consider for an optimal design: (1) Case 1 is $c = 1$ and $M = 4$; (2) Case 2 has $c = 1$, $M = 10$; (3) Case 3 will be $c = 2$ with $M = 4$; and (4) $c = 2$, $M = 10$ comprises Case 4. These are the values of c and M that we used in Chapter 13-16.

Here we use values for c, M and the fuzzy arrival (service) rates, to compute the $\overline{U}^{(N)},...,\overline{\Pi}^{(L)}$ and then find the values for $\overline{R},...,\overline{\Pi}$.

Now we pick values for the fuzzy arrival (service) rates. Some of the combinations employed here will be different from those used in Chapters 13-16. For example, in Chapter 13 we used $\overline{\lambda} = (3/4/5)$ ($\overline{\mu} = (5/6/7)$) for normal fuzzy arrival (service) rate but in this chapter we will employ a different pair of values for normal operating conditions. First we set $\overline{\lambda}^{(n)} = (4/5/6)$ and $\overline{\mu}^{(n)} = (5/6/7)$. Next $\overline{\lambda}^{(b)} = (6/7/8)$ and lastly $\overline{\lambda}^{(l)} = (3/4/5)$, $\overline{\mu}^{(l)} = (2/3/4)$.

All the needed fuzzy steady state probabilities have been determined in previous chapters. In fact, all the needed fuzzy numbers $\overline{U}^{(N)},...,\overline{R}^{(L)}$, without the $\overline{\Pi}$ values, were found in previous chapters. So now we need to

Case	(μ_1, μ_2, μ_3)	L_z	m_z	R_z
1	$(0.5, 0.2, 0.3)$	0.3518	1.0323	0.3784
	$(0.3, 0.5, 0.2)$	0.2667	0.9014	0.2781
	$(0.2, 0.3, 0.5)$	0.3043	1.1728	0.2347
2	$(0.5, 0.2, 0.3)$	1.0604	1.8844	1.2940
	$(0.3, 0.5, 0.2)$	0.7093	1.4707	0.8314
	$(0.2, 0.3, 0.5)$	0.7806	1.3801	0.6874
3	$(0.5, 0.2, 0.3)$	0.2128	0.6995	0.2885
	$(0.3, 0.5, 0.2)$	0.1609	0.5665	0.2872
	$(0.2, 0.3, 0.5)$	0.1713	1.0221	0.2107
4	$(0.5, 0.2, 0.3)$	0.3480	0.5556	0.4691
	$(0.3, 0.5, 0.2)$	0.2475	0.4753	0.3236
	$(0.2, 0.3, 0.5)$	0.2792	0.9573	0.2455

Table 17.4: Central Value and Certain Areas Under the Graph of $\overline{Z}$, the Four Cases in Example 17.3.1

find $\overline{\Pi}^{(W)}$, $W = N, B, L$ from equation (17.8). In equation (17.8) we use $\overline{Q} = (0.04/0.07/0.10)$, $\overline{C} = (0.27/0.30/0.33)$ and $\overline{T} = (0.83/0.95/1.07)$ as were employed in Chapter 14.

Having determined the fuzzy numbers for fuzzy profit we may next get the fuzzy numbers for $\overline{R}$, $\overline{U}$ and $\overline{\Pi}$ from equations (17.9) to (17.11). All fuzzy numbers are approximated through its alpha-cuts $\alpha = 0, 1/3, 2/3, 1$, so equations (17.9) to (17.11) are evaluated using interval arithmetic for $\alpha = 0, 1/3, 2/3, 1$.

The next step is to get $\overline{Z}$ from equation (17.12). In this calculation we used $K = 2$. In Chapters 13-16 we would have picked the value of the constant K so that the factor, now a fuzzy number, $[K - \overline{R}]$ is positive. In this example there are some situations were this fuzzy number is not positive. For example, in Case 2 with $(\mu_1, \mu_2, \mu_3) = (0.5, 0.2, 0.3)$. the right end point of the $\alpha = 0$ cut of $\overline{R}$ is 2.8822 so that the left end point of the $\alpha = 0$ cut of $[2 - \overline{R}]$ is -0.8622. However, negative values in the left end points of the alpha-cuts of $[2 - \overline{R}]$ are no problem in $max\overline{Z}$ because they just shift $\overline{Z}$ to the left, decrease its central value m_z and decrease the value of the objective function in equation (17.13). From $\overline{Z}$ we obtain L_z, m_z and R_z and these numbers are presented in Table 17.4 for various choices for the μ_i.

Using the results in Table 17.4 we may now easily find the values of the objective function, given in equation (17.13), for certain values of the λ_i. We used $K_1 = 1$ in these calculations. We see from Table 17.4 that $[1 - L_z]$ can be negative (Case 2, $(\mu_1, \mu_2, \mu_3) = (0.5, 0.2, 0.3)$). In previous chapters we chose K_1 so that the factor $[K_1 - L_z] > 0$. However, negative values for $[K_1 - L_z]$ are acceptable with the optimization problem in equation (17.13) because it just decreases the value of the objective function. Table 17.5 shows the values

Case	(μ_1, μ_2, μ_3)	$(1/3, 1/3, 1/3)$	$(0.2, 0.4, 0.4)$	$(0.2, 0.5, 0.3)$
1	$(0.5, 0.2, 0.3)$	0.6863	0.6939	0.7593
	$(0.3, 0.5, 0.2)$	0.6376	0.6185	0.6808
	$(0.2, 0.3, 0.5)$	0.7010	0.7021	0.7959
2	$(0.5, 0.2, 0.3)$	1.0393	1.2593	1.3183
	$(0.3, 0.5, 0.2)$	0.8642	0.9790	1.0429
	$(0.2, 0.3, 0.5)$	0.7623	0.8709	0.9402
3	$(0.5, 0.2, 0.3)$	0.5917	0.5526	0.5937
	$(0.3, 0.5, 0.2)$	0.5643	0.5093	0.5372
	$(0.2, 0.3, 0.5)$	0.6872	0.6588	0.7400
4	$(0.5, 0.2, 0.3)$	0.5589	0.5403	0.5489
	$(0.3, 0.5, 0.2)$	0.5171	0.4701	0.4852
	$(0.2, 0.3, 0.5)$	0.6412	0.6253	0.6965

Table 17.5: Values of the Objective Function in Example 17.3.1 for Certain Values of the Parameters μ_i and λ_i

of the objective function for three selections of the λ_i. The notation (a, b, c) at the top of this tables means $\lambda_1 = a$, $\lambda_2 = b$ and $\lambda_3 = c$.

For given values of the μ_i and λ_i, what is the optimal solution? Recall this is a maximization problem. The clear answer is Case 2 ($c = 1$, $M = 10$) in all cases.

Chapter 18

Summary and Future Research

18.1 Introduction

In this book we started with gathering data on the system and ended with optimization models to be used in web planning. We employed two methods in web planning: (1) fuzzy probabilities; and (2) fuzzy arrival rates and fuzzy service rates. So, we first present a brief summary of these two methods followed by our future research topics in this area. A summary of the computational algorithms used is in the next chapter. There is an increasing interest in web planning as seen by recent books in this area (see [4],[5],[6]) and new conferences on this topic (see [2],[3]).

18.2 Fuzzy Probabilities

We begin with obtaining data on the number of arrivals per unit time δ, and the number of service completions per unit time given that the server was being utilized at the start of the time interval. From this data we construct fuzzy numbers $\overline{p}(i)$ for the fuzzy probability that i customers arrive in time interval δ and $\overline{p}$ for the fuzzy probability that a customer leaves a server during time interval δ given that the customer was in the server at the start of the time interval. These fuzzy numbers are constructed by placing confidence intervals, one on top of the other, from a 99% confidence interval to a 0% confidence interval. The base of these fuzzy numbers, or the $\alpha = 0$ cut, is the 99% confidence interval.

From these fuzzy probabilities we determine α-cuts of the fuzzy transition probabilities in fuzzy transition matrices for Markov chains. The fuzzy transition matrices then produce α-cuts of the fuzzy steady state probabilities. The fuzzy steady state probabilities then determine α-cuts of the fuzzy system descriptors $\overline{U}$ = server utilization, $\overline{N}$ = average number of requests in the system, $\overline{X}$ = average server throughput , $\overline{R}$ = average response time, and $\overline{LC}$ = expected number of lost customers. The $\alpha = 0$ cut of these fuzzy numbers is like a 99% confidence interval for their value. For example, the spread (width) of the fuzzy number $\overline{R}$ shows the uncertainty in its value determined from the original data on the system. The α-cuts of $\overline{U}$, $\overline{N}$, $\overline{X}$, $\overline{R}$ and $\overline{LC}$ are then inputted to the optimization models for web planning.

The optimization models are in Chapters 13-17. By time time the reader gets to Chapter 17 we are sure he/she can make up other optimization models of interest to them. We hope that we have given enough details in the book so they can then use their own optimization models in web planning.

18.3 Fuzzy arrival/Service Rate

We begin with obtaining data on the number of arrivals per unit time and the time between service completions. From this data we construct fuzzy numbers $\overline{\lambda}$ for the fuzzy arrival rate and $\overline{\mu}$ for the fuzzy service rate. These fuzzy numbers are constructed by placing confidence intervals, one on top of the other, from a 99% confidence interval to a 0% confidence interval. The base of these fuzzy numbers, or the $\alpha = 0$ cut, is the 99% confidence interval.

From these fuzzy number $\overline{\lambda}$ and $\overline{\mu}$ we find α-cuts of the fuzzy steady state probabilities. The fuzzy steady state probabilities then determine α-cuts of the fuzzy system descriptors $\overline{U}$, $\overline{N}$, $\overline{X}$ and $\overline{R}$. The $\alpha = 0$ cut of these fuzzy numbers is like a 99% confidence interval for their value. For example, the spread (width) of the fuzzy number $\overline{U}$ shows the uncertainty in its value determined from the original data on the system. The α-cuts of $\overline{U}$, $\overline{N}$, $\overline{X}$ and $\overline{R}$ are then inputted to the optimization models for web planning.

18.4 Future Research

For the immediate future we see the following two areas of research.

18.4.1 CD-ROM

There is a lot of computation going from initial data to an optimization model. If there is a second edition of this book we would like to supply a CD-ROM to do most of these calculation. The ideal would be that the user inputs the system data, selects an optimization model and inputs the values of the needed parameters, and then sees the optimum solution.

18.4.2   Simulation

Another method of speeding up the calculations is to use a crisp (not fuzzy) simulation package using the α-cuts of the initial fuzzy numbers to get α-cuts of $\overline{U}$, $\overline{N}$, $\overline{X}$ and $\overline{R}$, see [1]. For example consider $\overline{\lambda}[0] = [\lambda_1, \lambda_2]$ and $\overline{\mu}[0] = [\mu_1, \mu_2]$. Randomly choose $\lambda \in [\lambda_1, \lambda_2]$ and $\mu \in [\mu_1, \mu_2]$ and then simulate the system using λ and μ. We will obtain a distribution of values for R (crisp R). From this distribution of values for R we obtain points in the $\alpha = 0$ cut of $\overline{R}$. Repeat the process to estimate the whole interval $\overline{R}[0]$.

18.5   References

1. J.J.Buckley, K.Reilly and X.Zheng: Crisp Simulation of Fuzzy Computations, Proc. ISUMA, Univ. Maryland, Sept. 21-24, 2003. To appear.

2. 2003 IEEE/WIC Int. Conf. on Web Intelligence (WI 2003), (www.comp.hkbu.edu.hk/WI03/).

3. Internet and Multimedia Systems and Applications (IMSA 2003),(www.iasted.org/conferences/2003/hawaii).

4. D.A.Menasce and V.A.F.Almeida: Capacity Planning for Web Performance, Prentice Hall, Upper Saddle River, N.J., 1998.

5. D.A.Menasce and V.A.F.Almeida: Scaling for E-Business: Technologies, Models, Performance and Capacity Planning, Prentice Hall, Upper Saddle River, N.J., 2000.

6. D.A.Menasce, V.A.F.Almeida: Capacity Planning for Web Services: Metrics, Models and Methods, Second Edition, Prentice Hall, Upper Saddle River, N.J., 2001.

Chapter 19

Computational Algorithms

19.1 Introduction

In this chapter we will review the sequence of computations needed in both approaches from the construction of the initial fuzzy numbers through the optimization models in Chapters 13-17. We start with the approach based on fuzzy probabilities followed by the method derived from fuzzy arrival rates and fuzzy service rates.

19.2 Computations: Fuzzy Probabilities

The method begins with obtaining $\overline{p}(i) = $ the fuzzy probability of i customers arriving in time interval δ and getting $\overline{p} = $ the fuzzy probability that a customer leaves a server during time period δ given that the customer was in the server at the start of the time interval. The details on constructing these triangular shaped fuzzy numbers was presented in Section 3.2 of Chapter 3. Obtaining fuzzy numbers from expert opinion was discussed in Section 3.4.

Using the $\overline{p}(i)$ and $\overline{p}$ we next need to compute the fuzzy transition probabilities $\overline{p}_{ij}$ in the fuzzy transition matrix $\overline{P}$ for a fuzzy Markov chain. Actually we only want some α-cuts of the $\overline{p}_{ij}$ in $\overline{P}$. These optimization problems were first discussed in Sections 3.5 and 3.6 of Chapter 3. See also "Step 1" in Sections 6.2 and 8.2 and "Step 1" in Examples 13.2.1.1, 14.2.1 for further details. Basically the solution methods were: (1) use "common sense" for the easy problems; (2) try elementary calculus (if the partial derivative with respect to a certain variable is positive, the the objective function is an increasing function of that variable); (3) it is a linear programming problem and then use "simplex" in Maple [2]: and (4) it is a non-linear programming problem and then use the Premium Solver Platform V5.0 from Frontline Systems [1].

From $\overline{P}$ we now need to find the fuzzy steady state probabilities. Chapter 4 presents the three types of fuzzy Markov chains encountered in this book and discusses possible solutions strategies to get these fuzzy steady state probabilities. In this book $\overline{P}$ is either 5×5 when $M = 4$, or it is 11×11 when $M = 10$. M is system capacity. In general our solution strategy was: (1) use the Premium Solver Platform V5.0 for all 5×5 $\overline{P}$ and sometimes also employ our genetic algorithm; and (2) use only the genetic algorithm on the 11×11 $\overline{P}$. The reason for not employing the Premium Solver Platform on the 11×11 matrices will be discussed below. We will first present the general optimization problem for the Premium Solver Platform. Then we go through the structure of our genetic algorithm.

19.2.1　Premium Solver Problem

Let $\overline{P}$ be $n \times n$ and $\overline{P} = (\overline{p}_{ij})$. Define $\overline{p}_{ij}[\alpha] = [p_{ij1}(\alpha), p_{ij2}(\alpha)]$ all i, j and $0 \leq \alpha \leq 1$. Next let $w = (w_1, ..., w_n)$ and $\overline{w} = (\overline{w}_1, ..., \overline{w}_n)$. Recall, see Chapter 4, if P is the crisp transition matrix for a crisp, regular, Markov chain, then $wP = w$, $0 \leq w_i \leq 1$ all i and $w_1 + ... + w_n = 1$ will give the steady state probabilities w_i. Define $\overline{w}_i[\alpha] = [w_{i1}(\alpha), w_{i2}(\alpha)]$ all i and all α.

For selected values of α we wish to solve

$$min[w_i] = w_{i1}(\alpha), \tag{19.1}$$

and

$$max[w_i] = w_{i2}(\alpha), \tag{19.2}$$

for $i = 1, ..., n$. The constraints are

$$p_{ij1}(\alpha) \leq p_{ij} \leq p_{ij2}(\alpha), \tag{19.3}$$

for all i, j and

$$p_{i1} + ... + p_{in} = 1, \tag{19.4}$$

for $i = 1, ..., n$. Set the crisp transition matrix $P = (p_{ij})$. Notice that all row sums equal one. Also

$$wP = w, \quad 0 \leq w_i \leq 1, \quad w_1 + ... + w_n = 1. \tag{19.5}$$

In this optimization problem the variables are $p_{11}, ..., p_{nn}, w_1, ..., w_n$. A feasible solution is any values of the variables that satisfies all the constraints. Let $\mathcal{F}$ be the set of all feasible solutions.

Here we have a slight notational problem. Let M be system capacity (in the queue and in the servers). In this section we are numbering the rows/columns in the transition matrices $1, 2, ..., M + 1$ with $n = M + 1$. In the book we usually number the rows/columns of the transition matrices as $0, 1, 2, ..., M$.

When P is 5×5, or $M = 4$, the Premium Solver Platform solved this optimization problem easily. We used the multiple start feature where the algorithm selects multiple feasible solutions in $\mathcal{F}$ to initiate the process and them selects the best result. When P was 11×11, or $M = 10$, the Premium Solver Platform could not solve the optimization problem because it could not find a feasible solution in $\mathcal{F}$ to get started. Apparently, in this case the feasible set $\mathcal{F}$ must have a very irregular, and possibly very "thin", shape. We can find feasible solutions in $\mathcal{F}$ by hand calculations so the feasible set $\mathcal{F}$ is non-empty. It is possible to feed the algorithm a starting feasible solution, but with multiple starts we would need to give it lots of starting feasible solutions, for each new fuzzy transition matrix. But, in the 11×11 case there are 132 variables whose values need to be entered for each of the multiple starts and for each separate problem to be solved. Hence, we did not use this method for the 11×11 fuzzy transition matrices.

19.2.2 Genetic Algorithm

We will use the same notation as in the previous subsection. Let us first describe the feasible set $\mathcal{F}$ for the genetic algorithm. Let $v = (p_{11}, p_{12}, ..., p_{nn})$ a $1 \times n^2$ vector of all the probabilities in a $n \times n$ transition matrix $P = (p_{ij})$ for a regular Markov chain. Now $\mathcal{F}$ consists of all v so that

$$p_{ij1}(\alpha) \leq p_{ij} \leq p_{ij2}(\alpha), \tag{19.6}$$

for all i, j and

$$p_{i1} + ... + p_{in} = 1, \tag{19.7}$$

for $i = 1, ..., n$ (all the row sums are one). It is important that this $\mathcal{F}$ is convex. What this means is that if v^a and v^b are in $\mathcal{F}$, then so is v^c where

$$v^c = \lambda v^a + (1 - \lambda)v^b, \tag{19.8}$$

for all $0 \leq \lambda \leq 1$. This fact will be used in the crossover operation in the genetic algorithm. The feasible set now is different from that in the previous subsection 19.2.1.

What we need to describe is the initial population $\mathcal{P}_0$, the fitness function, crossover, mutation and the next generation $\mathcal{P}_1$. The initial population $\mathcal{P}_0$ is just a set of randomly generated 100 $v_i \in \mathcal{F}$. To describe the construction of the next population let $v \in \mathcal{P}_0$ and using this v construct the transition matrix $P = (p_{ij})$. Next compute the vector $w = (w_1, ..., w_n)$ so that

$$wP = w, \quad 0 \leq w_i \leq 1, \quad w_1 + ... + w_n = 1. \tag{19.9}$$

Let $w^{(i)} = (w_1^{(i)}, ..., w_n^{(i)})$, $1 \leq i \leq 100$, be all the vectors w obtained from equation (19.9) using all the $v \in \mathcal{P}_0$. Suppose in this run of the genetic algorithm we are looking for $w_{31}(\alpha)$ the left end point of the interval $w_3[\alpha]$. We now sort the $w_3^{(i)}$, $1 \leq i \leq 100$, from smallest to largest.

The crossover operation generates possibly a new population member from two elements in $\mathcal{P}_0$. We will randomly choose two members of $\mathcal{P}_0$ for crossover. But first we randomly choose two w from the set $w^{(i)}$, $1 \le i \le 100$. In this random process it will be more likely that we pick a $w^{(i)}$ having a smaller value of $w_3^{(i)}$ than from those having the larger values of $w_3^{(i)}$ (this is easily accomplished from the sorting described above). Suppose we picked w^a and w^b. These two then correspond to v^a and v^b in $\mathcal{P}_0$. From v^a and v^b we determine $v^c \in \mathcal{F}$ as follows:

$$v^c = \lambda v^a + (1 - \lambda)v^b. \tag{19.10}$$

This v^c is in $\mathcal{F}$ for any $\lambda \in [0, 1]$. A value of λ is randomly generated to get v^c. Generate around 10 of the v^c in this manner and put them all in $\mathcal{P}_0$. Calculate their corresponding vectors w. Notice that $\mathcal{P}_0$ has grown to be more than 100 members.

Next we discuss mutation. To do this we need to consider the following equation:

$$1 - p_{in2}(\alpha) \le \sum_{j=1}^{n-1} p_{ij} \le 1 - p_{in1}(\alpha), \tag{19.11}$$

for $i = 1, 2, ..., n$. We next randomly choose a few (maybe 5) v from $\mathcal{P}_0$ for mutation. Suppose $v = (p_{11}, ..., p_{nn})$ was chosen. Randomly choose an element in v which is not p_{in}, $1 \le i \le n$ (not the end of a row in P). Assume we picked p_{46}. Now

$$p_{461}(\alpha) \le p_{46} \le p_{462}(\alpha). \tag{19.12}$$

Randomly choose a value in this interval $[p_{461}(\alpha), p_{462}(\alpha)]$. Assume we got p_{46}^*. If equation (19.11) is satisfied for this $p_{46}^* = p_{46}$ we keep it, otherwise it is discarded and we randomly choose another value in the interval. So assume that we keep this p_{46}^*. Substitute p_{46}^* for p_{46} in v. Now adjust the value of p_{4n} so that the fourth row sum is one (we may always do this since equation (19.11) was satisfied). We now have a new (mutated) v in $\mathcal{P}_0$. After doing this maybe five times we have introduced new mutated members into $\mathcal{P}_0$. For all the mutated v compute their corresponding vectors w, discard the old values w.

Now sort all the $w_3^{(i)}$ from smallest to largest for all the v in $\mathcal{P}_0$ and choose the 100 smallest. Determine the v in $\mathcal{P}_0$ corresponding to the 100 smallest $w_3^{(i)}$, these v are the next generation $\mathcal{P}_1$.

Continue through this process of getting $\mathcal{P}_i$, calculating the vectors w, sorting, crossover and mutation for Θ generations. We usually used 200 generations. Then we will have a good estimate of $w_{31}(\alpha)$. This is repeated to estimate all the end points of the alpha-cuts of the fuzzy steady state probabilities.

In our initial applications of the genetic algorithm we experimented with different sizes for $\mathcal{P}_0$ and different values for Θ. For example: (1) in Table 7.2 $\mathcal{P}_0$ had 200 members and $\Theta = 100,000$; and (2) for Table 9.2 we had 200 in $\mathcal{P}_0$ and $\Theta = 7,000$. However, when we got to Chapters 13-17 we always used 100 for the size of the initial population and had values between 200 and 700 for Θ.

In our previous experience with genetic algorithms we noticed that too often crossover produces a result not in the feasible set. In fact the algorithm can spend too much time discarding the results of crossover because they are not in $\mathcal{F}$. The same result may occur in mutation. Since $\mathcal{F}$ is convex our crossover always gives a result in the feasible set.

Looking at the optimal solutions, especially the 5×5 case, we found that the optimal v was quite often on the boundary of $\mathcal{F}$. The solution v is on the boundary of $\mathcal{F}$ when one, or more, of the inequalities in equation (19.6) is an equality. So in the genetic algorithm we need population members $v \in \mathcal{P}_i$ on the boundary of $\mathcal{F}$. Notice that crossover, equation (19.10), always gives a v^c "between" v^a and v^b. So employing only crossover new populations will tend to migrate to the "center" of $\mathcal{F}$. Hence, the important operation of mutation is to make some v exist on, or near, the boundary of $\mathcal{F}$. We can get v from mutation on, or near, the boundary of $\mathcal{F}$ by having it more likely that the p_{ij} we choose in $[p_{ij1}(\alpha), p_{ij2}(\alpha)]$ is at, or near, the end points of the interval.

The author wishes to thank Mr. Xidong Zheng (Dept. of Computer and Information Sciences, University of Alabama at Birmingham, Birmingham, Alabama, 35294) for the design and implementation of the genetic algorithm used in this book.

Using the fuzzy steady state probabilities the next thing to compute are the alpha-cuts of the system descriptors $\overline{U}$, $\overline{R}$, $\overline{X}$ and $\overline{R}$. Finding an α-cut of $\overline{U}$, $\overline{N}$ and $\overline{X}$ is to solve a linear programming problem. We solved these linear programming problems using "simplex" in Maple [2]. $\overline{R}$ is just $\overline{N}/\overline{X}$.

19.2.3 Optimization Models

The optimization models are in Chapters 13-17. All of these optimization models are computationally easy because they are discrete optimization models. What we mean by discrete is that the variables can take on only a finite number of values. For example, in Chapter 14: (1) there is only one or two servers ($c = 1, 2$); (2) there are only two kinds of system capacity ($M = 4, 10$); (3)there are only two types of fuzzy arrival probabilities; and (4) there are two types of fuzzy service rates. This gives us a total of sixteen cases to look at to find the optimal solution.

For all of these optimization models we wrote a simple program in Maple [2] where you input the values of the variables and then Maple will compute the value of the objective function. In Chapter 14, for each objective function, we have to compare only sixteen numbers to see the optimal solution.

When we were using the method of ordering the fuzzy numbers from smallest to largest (Section 2.6 of Chapter 2) we wrote a simple Maple [2] program to evaluate equation (2.42).

19.3 Computations: Fuzzy Arrivals and Service Rates

The method begins with obtaining $\overline{\lambda}$ = the fuzzy arrival rate and getting $\overline{\mu}$ = the fuzzy service rate. The details on constructing these triangular shaped fuzzy numbers from a set of confidence intervals was presented in Section 3.3 of Chapter 3. Obtaining fuzzy numbers from expert opinion was discussed in Section 3.4.

Using the $\overline{\lambda}$ and $\overline{\mu}$ we next need to compute the fuzzy steady state probabilities. This computation is much easier then for the fuzzy probabilities approach. We simply fuzzified the equations from classical queuing theory. This is discussed in Chapter 12. In Chapter 12 we only did the $\alpha = 0, 1$ cuts but for the optimization models in Chapters 13-17 we used the $\alpha = 0, 1/3, 2/3, 1$ cuts of the fuzzy steady state probabilities. Finding these alpha-cuts are all solutions to a non-linear optimization problem. After Chapter 12 all of these non-linear optimization problems were solved using the Premium Solver Platform V5.0 from Frontline Systems [1] employing the multiple start option.

Using the fuzzy steady state probabilities the next things to compute are the alpha-cuts of the system descriptors $\overline{U}$, $\overline{R}$, $\overline{X}$ and $\overline{R}$. Finding an α-cut of $\overline{U}$, $\overline{N}$ and $\overline{X}$ is to solve a linear programming problem. We solved these linear programming problems using "simplex" in Maple [2]. $\overline{R}$ is just $\overline{N}/\overline{X}$. Here we obtained the $\alpha = 0, 1/3, 2/3, 1$-cuts of these fuzzy numbers.

The computations in the optimization models is the same, except we now work with more α-cuts, as that described for the procedure based on fuzzy probabilities.

19.4 References

1. Frontline Systems (www.frontsys.com).

2. Maple 6, Waterloo Maple Inc., Waterloo, Canada.

Index

List of Figures

List of Tables